Alzheimer's Turning Point

Jack C. de la Torre

Alzheimer's Turning Point

A Vascular Approach to Clinical Prevention

Springer

Jack C. de la Torre, MD, PhD
Professor of Neuropsychology (Adjunct)
Department of Psychology
University of Texas at Austin
Austin, TX
USA

ISBN 978-3-319-34056-2 ISBN 978-3-319-34057-9 (eBook)
DOI 10.1007/978-3-319-34057-9

Library of Congress Control Number: 2016939367

Printed on acid-free paper

This Springer imprint is published by Springer Nature
The registered company is Springer International Publishing AG Switzerland

To Lauren, who lights up my heart even when skies are gray.

Dad

Foreword

There are many words in the medical language beginning with the Latin prefix "De". Most of them implicate a metaphor, such as "delirium"—deviating ("de"viating) from the binary or depression—losing pressure. Dementia, however, is a term clearly pointing at the loss of a critical part of human existence, and it is not a metaphor. Dementia is a condition already described in most ancient times to depict the loss of ability to recognize, acknowledge, and judge. As this feature of a pathologic condition affecting the soul, the spirit, the "nous", the intellect, has remained quite stable over the centuries, it has not been deemed necessary by the medical community to somehow take a closer look at the term "dementia" even after Alois Alzheimer's publication of Auguste Detre's mental state condition. Dementia is a complex and complicated topic, and the approach to it is being rendered difficult directly at the linguistic level when mentioning it. One may ask, what is dementia linguistically? What is it semantically?

In his book, Dr. de la Torre summarizes the Alzheimer's disease evidence and knowledge about one of the most impressive clinical failures in the history of medicine. Interestingly, despite this extremely negative continuum, the readers will appreciate not to be persuaded, even in a single sentence throughout the text, to have actively contributed to the Alzheimer's clinical failure.

Rather, the whole body of information presented in this text is permeated by a positive, pragmatic attitude to solve the problem of this devastating disease. Dr. de la Torre's work opens the door to a new management strategy to prevent memory disturbances and cognitive impairment which are by and large linked to increasing age. The three characteristics of this "new" approach are as follows: (1) It is based on a precise rationale built from a large body of empirical scientific evidence; (2) it has been systematically assembled for the first time in this text; (3) its reasoning is derived from information virtually under the sight of every health professional in the field of cognitive impairment for decades.

The fact that there is no cure for dementia is because dementia is a multifactorial, geriatric syndrome lacking the prerequisite "single bullet" pathophysiological mechanism; this fact excellently reflects two postulates which currently are themselves turning into facts. The first postulate is that every additional vascular "hit" further lowering cerebral perfusion in a brain already burdened by age-related hypoperfusion will eventually lead to a cascade of pathological events ultimately precipitating neurodegeneration and clinically overt cognitive impairment. The second postulate is that there is no established algorithm to characterize the general level of brain perfusion susceptible of reaching a pathological threshold following an additional vascular "hit." A critical threshold of cerebral blood flow might be attained in an elderly individual at a very advanced age from one single vascular risk factor, while in another older individual of a lesser age, the critical brain perfusion threshold might be reached after multiple cardiovascular insults. In this sense, all the nuances between these extremes are possible.

In the meantime, most studies showing the positive effects of lifestyle habits against dementia onset are confirming that healthy behaviors typically improving cardiovascular performance will delay the onset of dementia. On the other side of the coin, and for those who tend to look at the not so strikingly positive effects of such behaviors, these results are likely to be negatively impacted by the inherent epidemiological fault of the non-personalized, non-state-of-the-art intervention.

In this book, Dr. de la Torre advocates what the experts in anti-aging medicine and geriatrics should already know and should already be doing: acknowledge the heterogeneity of Alzheimer's disease and personalize the management approach to each patient in the interest of preventing whenever possible the risk or onset of severe cognitive impairment.

The author in his text not only shows a successful approach to cognitive impairment management but also indicates the scientific rationale and methodology for its implementation. Reading this book, the impact of cognitive impairment prevention through targeted, personalized vascular risk control becomes evident and anagously discloses its relevant socioeconomic implication, that is, systematically preventing the "millennium epidemic" of an Alzheimer's socioeconomic disaster that would render its future management unsustainable.

This is in contrast with the current illusory target of a pharmaceutical cure that ignores the clinical struggle with patient care and the consequences of random disease management. Not applying a comprehensive assessment of the adult and aging population to identify modifiable vascular and lifestyle risk factors that can influence the onset and course of dementia seems aimless. Currently applying usual care too late in the disease while waiting for utopic curative drug development to treat Alzheimer's dementia is unethical and insensitive. This book provides ample evidence to correct these ineffective courses of action.

March 2016

Prof. Maria Cristina Polidori MD, PhD, FRCP
Geriatrician, Gerontologist, Head
Unit for Ageing Clinical Research
Department of Medicine II
University Hospital of Cologne, Germany

Preface

Clinical progress in Alzheimer's disease has followed a tired, tenuous, and trackless course for the last 25 years. This medical failure to provide even adequate control of patients at risk of dementia needs to change quickly to avoid a socioeconomic and medical catastrophe in the near future.

Now, a turning point that can revitalize our thinking on how to markedly control Alzheimer's disease is within our grasp. This turning point promises a new beginning about how Alzheimer's should be medically managed to prevent, reverse, or delay onset of this mind-shattering disease. The essence of containing Alzheimer's is not simply to diagnose the disease onset, provide some pills, and send the patient home, which is generally the present practice. *The main goal of managing this incurable dementia is by prevention.* Alzheimer's prevention is now achievable as discussed in the chapters that follow. A primary step in the prevention process is early detection of patients at risk of acquiring Alzheimer's using state-of-the-art diagnostics and application of a personalized clinical plan of action. This is best achieved by the establishment of Heart–Brain Clinics dedicated to containing what is now the most important medical challenge of the twenty-first century. This clinical approach will not eradicate Alzheimer's but as Heart–Brain Clinics proliferate nationally and abroad, a significant reduction in the incidence of new Alzheimer's cases will be seen to provide a significant grace period of healthy aging to many who would otherwise become easy victims of this dementia.

This book comprehensibly discusses these and other relevant issues in clear, descriptive language and focuses on the actions that will determine how this turning point can be rapidly implemented. The turning point in Alzheimer's disease is certain to impact the individual who lives long enough to develop a risk to this dementia.

Alzheimer's disease, Alzheimer's dementia or Alzheimer's for short, is now considered the most common and irreversible destroyer of the human brain. Unlike some infectious diseases, there is no magic bullet to cure it once it is diagnosed. If Alzheimer's is not reasonably contained, it will rise ballistically from a present 45 to 135 million people affected worldwide by 2050. In the same time period, that number will triple in the USA, from 5.5 to nearly 14 million people affected.

Every year in the USA, Alzheimer's claims more victims after age 65 than cancer, heart disease, and AIDS combined, yet the government spends less money on Alzheimer's research than all three of these disorders. A paradoxical fact is that while deaths from cancer, heart disease, and AIDS have plummeted in recent years, Alzheimer's deaths keep climbing. In the USA, the annual cost of Alzheimer's to society is estimated to be $226 billion, a cost that will rise exponentially by 2050 to over $1 trillion with devastating consequences on Medicare, Medicaid, and the US healthcare system.

A book like this cannot do justice to the vast and complex field of Alzheimer's research. However, an attempt is made to focus on the importance of where we have been for the last 25 years with respect to our present view of Alzheimer's disease. Looking back at recent past failures will provide a crucial clue as to where we need to go from this point on to provide real hope to those who may be vulnerable of acquiring Alzheimer's.

This book is a departure from most others on the market because it does not try to provide advice on how to cope with Alzheimer's or provide suggestions regarding what foods to eat or what activities to take up to reduce the risk of dementia. Dozens of books on the market are available that provide such information. Instead, this book is a practical guide to a plan on how the number of new Alzheimer's cases recorded every year can be substantially reduced using evidence-based medical information derived from a series of recent clinical studies. These clinical studies are melded with the author's extensive experience in laboratory and clinical research of Alzheimer's disease gained in the last 25 years as a professor of neurosurgery, neuroscience, and pharmacology at several North American medical schools.

As a continuous blueprint, each chapter pieces together an essential link that collectively supports the final conclusion that Alzheimer's can be successfully managed by the application of new technology, state-of-the-art diagnostic tools, and recently gained knowledge of what really causes dementia and what does not. This book tries to minimize scientific jargon whenever possible, and when technical terms are unavoidable, descriptive explanations follow to make plain the idea or concept. The compiled information gathered here will be helpful to physicians in identifying the preclinical risk factors associated with the earliest stages of cognitive decline, an approach that can help the patient delay or avoid the Alzheimer's bullet. This book should also be useful to graduate, postdoctoral, and established neuroscientists who have an interest in neurodegenerative diseases.

On a lighter side, the book will show how laboratory experiments using young and old rats swimming in a water tank in the author's laboratory provided pivotal information leading to the possible cause of Alzheimer's and how these findings helped identify more than two dozen risk factors to this dementia. *The identification of Alzheimer's vascular risk factors is probably one of the most important findings made in the history of this disease because first, virtually every acquired risk factor has a vascular component that can be therapeutically targeted to prevent its consequences and, second, because these vascular risk factors provide a key to unlocking the mystery that has shrouded Alzheimer's pathology for over a century.* That key has to do with the fact that risk factors to Alzheimer's share a crucial common denominator. That is, *virtually all Alzheimer's risk factors presently known are able to reduce blood flow to the brain.* The book discusses how poor blood flow to the brain in elderly people will commonly transform a healthy brain into one that eventually expresses cognitive impairment and severe memory loss. Such a process can take decades to form. Could poor blood flow to the brain have triggered President Ronald Reagan's Alzheimer's disease following his assassination attempt in 1981? The reader will see how this event may have evolved.

The chapters that follow will explore why clinical research in Alzheimer's disease has been a trackless waste for the last quarter century. Knowing *why* we failed for over a century to manage a lethal medical disorder such as Alzheimer's is the first step in determining *how* to construct a plan that can address and resolve prior clinical fiascos.

Clinical research is defined as the science that searches to find treatments, interventions, or management to either cure, prevent, or ease medical problems in humans. It uses clinical trials to examine the efficacy and safety of whatever treatment or intervention is being tested. My book discusses why dozens of clinical trials at a cost of hundreds of millions of dollars have ended in total failure thanks largely to a greedy and corrupt pharmaceutical industry. Moreover, it will be seen how bad hypotheses delay and often obstruct scientific progress, why the government remains idle to its most colossal medical problem, and why the pharmaceutical industry capitalizes on the despondence and hopelessness of victims and caretakers of this dementia to push their dubious nostrums that often do more harm than good on those who use them.

All is not gloom, however. A *turning point for Alzheimer's* is attainable, and it is hoped this book will accelerate that process. This turning point is no rabbit being pulled out of a magician's hat but a carefully constructed stratagem to shift our thinking from its present clinical quagmire and apply a rational program to directly control the staggering rise of new Alzheimer's cases expected to triple by 2050.

After reading this book, I am confident that the reader will feel sufficiently informed to disagree with the Doris Day's popular song that nihilistically claims "whatever will be, will be." Instead, the reader may realize that fate is often in the hands of the individual who refuses to accept a doomsday philosophy. This existentialist approach is artfully reflected in Shakespeare's tragedy Julius Caesar when Cassius remarks to Brutus, "The fault, dear Brutus, is not in our stars, but in ourselves."

For reasons that will become clear, we will examine in the chapters ahead the seamless logic of this Shakespearian penchant with respect to what an individual can do with the help of good medical practice to slowdown, stop, or reverse the clinical signs that may forecast the ominous beginnings of Alzheimer's dementia.

This book is highly recommended to a wide audience of readers consisting of researchers and clinicians with an interest in neurodegeneration and, particularly, Alzheimer's disease.

Contents

About the Author

Jack C. de la Torre MD, PhD began his research studies of Alzheimer's disease in 1990. The first series of experiments he conducted were based on his development of an aged rat model whose cerebral blood flow could be manipulated peripherally by occluding two or three of the four arteries supplying the brain. In dozens of experiments in his laboratory, this rodent model provided crucial information concerning the effects of chronic brain hypoperfusion on memory impairment, learning acquisition, neurochemical abnormalities, selective neuronal damage, and progressive neurodegeneration. These physiopathological changes in the aged rat model characterize the preclinical stage that precedes Alzheimer's onset.

Findings obtained from the brain hypoperfusion rat model led Dr. de la Torre to propose in 1993 that Alzheimer's is a vascular disorder with neurodegenerative consequences, a concept that challenged the prevailing Abeta hypothesis as the cause of Alzheimer's disease. The vascular hypothesis of Alzheimer's disease as he proposed has become a mother lode of basic research and clinical studies.

Dr. de la Torre began his graduate education in neuroanatomy at the University of Chicago, and he finished his Ph.D. in neuroanatomy at the University of Geneva. His doctoral research thesis involved his discovery that combining L-dopa with a dopa decarboxylase inhibitor (DDCI) could breach the blood–brain barrier in rats allowing a substantial increase in brain dopamine levels. This drug combination prompted his thesis advisor, neurologist Dr. René Tissot, to successfully treat severe Parkinson's disease patients and virtually reverse their neurological deficits. The Swiss pharmaceutical Hoffmann-La Roche then quickly developed L-dopa and DDCI into a pill form called Madopar which became a primary treatment for Parkinsonism. Post-doctoral work by Dr. de la Torre in neuropathology followed at the University of Geneva Bel-Air Psychiatric Clinic and in neurophysiology at the University of Zurich Institute of Physiology.

Dr. de la Torre obtained his MD after attending the University of Madrid Faculty of Medicine (Spain) and the Autonomous University of Ciudad Juárez (Mexico). Graduate medical training in neuropathology followed at the University of Miami Hospitals and Clinics.

Dr. de la Torre has written over 200 peer-reviewed articles and edited or coedited ten volumes on the vascular pathophysiology of Alzheimer's disease.

He is the author of Dynamics of Brain Monoamines and the translator of Ramón y Cajal's classic tome The Neuron and the Glial Cell. He is presently on the editorial board of several Alzheimer's journals and is senior editor in the Journal of Alzheimer's Disease.

Dr. de la Torre has held professorial appointments in neurosurgery and neuroscience departments at the University of Chicago, Northwestern University, University of Ottawa, and Case Western Reserve University. His present appointment is in the Department of Psychology at the University of Texas, Austin, where he is Professor of Psychology (adjunct).

Chapter 1
How Does Alzheimer's Begin and Who Gets It?

Alzheimer's disease, Alzheimer's dementia or Alzheimer's for short, is one of the grimmest diagnoses a physician can deliver to a patient. It conveys not only an unfaltering death sentence but an extended period of suffering, humiliation, and confusion before dying. In the USA, a person is diagnosed with this dementia every 67 s, and by 2050, it will be every 33 s. No one is sure what causes it, and as with any disease, knowing the cause of Alzheimer's is the key to containing it. This book attempts to shine some light into the possible cause of Alzheimer's and how this dementia can be prevented.

There is no cure for Alzheimer's, and there is no hope of recovery. The only hope is to **avoid the conditions that can promote or lead to Alzheimer's**. These preclinical conditions are called **risk factors,** and good progress has been made in the last few years to identify, detect, and, in some cases, prevent these risk factors from converting to Alzheimer's disease. Much more needs to be done and quickly. Risk factors probably constitute the most important targets in avoiding the disease altogether because virtually all risk factors known so far are vascular-related. The main vascular risk factors for Alzheimer's and what can be done about them are fully discussed in Chap. 12.

How can we define dementia? The present-day snappy definition, according to the Alzheimer's Association, is that of "**a neurodegenerative dementia that causes problems with memory, thinking, and behavior. Symptoms usually develop slowly and get worse over time, becoming severe enough to interfere with daily tasks and functional survival**" [1].

The problem with this definition is that it could be applied to a number of other dementias whose main symptoms are also memory loss, faulty cognition, and behavioral changes occurring over time (see Sect. 6).

By contrast, **our research has led us to believe that Alzheimer's is a vascular disorder with neurodegenerative consequences**, not the other way around as is classically believed [2, 3]. This view is a departure from previous understanding and can be described as follows. **Alzheimer's disease is a vascular disorder of the**

J.C. de la Torre, *Alzheimer's Turning Point*,
DOI 10.1007/978-3-319-34057-9_1

elderly brain caused by chronic cerebral hypoperfusion leading to neurodegenerative damage, progressive cognitive decline, and death.

With this latter definition, the triggering cause of the disease is recognized (cerebral hypoperfusion), its consequences on the brain are anticipated (neurodegeneration of brain cells), and the eventual mental outcome (cognitive loss) is predicted. This definition has not been pulled out of a hat by us since there is considerable evidence to support its validity as will be seen in the chapters that follow. What seems important about a legitimate definition of Alzheimer's or any other disease is that it should allow the most appropriate targeting of treatments designed to address the defining terms of the disorder. This is one of the reasons why the vascular hypothesis definition is clinically useful within the framework of dementia prevention.

At the present time, a consensus of investigators recognizes two forms of Alzheimer's. The most common form which affects more than 95 % of all the people affected is called ***sporadic* or late-onset Alzheimer's disease**. No particular gene seems responsible for this form of Alzheimer's, and patients over age 60 are more prone to develop the disease. The second form is called **early-onset or *familial* Alzheimer's disease** and is very rare, affecting a small number of families from the mutation of 3 genes, PS1, PS2, and amyloid precursor protein (APP). The familial form generally affects less than 5 % of people with Alzheimer's and usually strikes before age 60, sometimes as early as age 30–50. The familial form is autosomal dominant and can be inherited from either parent carrying the mutated gene. There is therefore a 50 % chance of getting the disease when the mutated gene for Alzheimer's disease is carried by one parent.

The APP gene is located on the long arm of chromosome 21 and was the first gene implicated in Alzheimer's disease. At this moment, about 15 disease-causing mutations have been linked to APP in which familial Alzheimer's accounts for about 5 % with most individuals developing symptoms between the ages of 40–60.

The reason for the genetic implication in familial Alzheimer's is because APP gene mutation also causes Alzheimer's in Down syndrome. These are patients who carry 3 copies of chromosome 21. The normal *APP* gene that everyone carries provides instructions for making the APP protein. This protein is normally found in many tissues and organs, including the brain and spinal cord, but its exact function in the body is not yet clear. When APP mutates, an increased amount of Abeta peptide forms in the brain creating clumps of senile plaques that can be seen in many Alzheimer's brains but also in healthy, elderly people with normal cognition [4].

Little evidence indicates that APP is overexpressed in the Alzheimer's brain as some investigators have claimed despite considerable research evidence that have questioned this notion [5]. More strikingly, most Alzheimer's mouse models do not show substantial neuronal loss, despite the presence of large depositions of amyloid [2, 6]. Further, in contrast to human Alzheimer's where loss of synapses is substantial, Alzheimer's mice models show a highly variable presentation. Some mice show increased synaptic density in specific brain regions, while other mice models show a loss of synaptic density. It is estimated that 100 trillion synapses exist in the human brain connecting about 100 billion neurons. It is not known how many of these synapses and neurons need to die in cognitive-related regions to create an Alzheimer's state.

In the event the APP mutation can be prevented someday by genetic manipulation, it will represent a cure for familial Alzheimer's but only in a small population of patients that inherit the mutated APP gene. The possibility of a cure for non-genetic Alzheimer's is unlikely at this time (see Chap. 13)

A large percentage of cognitively normal individuals have been found to harbor numerous degenerative changes that meet the criteria for an Alzheimer's diagnosis while still living or at autopsy [4]. This pathologic discrepancy has prompted some investigators to suggest that Abeta formation in the brain is a protective reaction to neurodegeneration [6]. This interpretation is supported by numerous studies showing that **the deposition of Abeta-containing plaques does not correlate with neuropathology, loss of neurons, or cognitive impairment.** Moreover, the total removal of Abeta-containing plaques from the brain of Alzheimer's patients does not improve their cognition when anti-Abeta vaccines were given in clinical trials [7]. In fact, some patients developed encephalitis or brain swelling after receiving these treatments.

This paradox has created a small problem for investigators who advocate Abeta-containing plaques as the cause of Alzheimer's and has given rise to the suggestion that cognitively normal individuals with a high plaque burden are in a preclinical stage of Alzheimer's not yet expressed [8]. However, there is no indication that this suggestion is valid when longitudinal follow-up studies of these plaque-burdened individuals are reviewed [9]. **Moreover, a meta-analytic survey of the literature shows that increased Abeta accumulation in the brain has a minor effect on cognitive performance, especially memory ability, thus questioning the role of Abeta as either a causal factor in Alzheimer's or a 'toxic' molecule to neurons** [10].

It has been shown that these Abeta deposits in Alzheimer's brain generally reach a plateau where their numbers do not increase despite declining cognition [11] (see Chap. 5, **A die-hard theory**). The enveloping Abeta shadow over Alzheimer's research nevertheless continues to this day with stubborn support from the pharmaceutical industry that has a vested interest in marketing anti-Abeta therapy.

It should be emphasized that if the APP-mutated gene is not inherited, the Alzheimer's familial form will not develop from the normal APP gene. Members of a family who do not inherit the gene mutation are no more likely to get Alzheimer's than are other members of the general population. About 95 % of people living In North America and Europe who are aged 65 and older are affected by the non-genetic type of dementia, so-called sporadic Alzheimer's [12]. This rate can double every 5 years after age 60 [13].

Throughout this book, an evidence-based foundation will be presented that reinforces our pathophysiologic (rather than genetic) definition of Alzheimer's while providing important new clues of the complex biological abnormalities associated with cognitive deterioration. This evidence-based foundation derives from epidemiological, pharmacological, pathological, clinical, and behavioral findings that link poor blood flow to the brain during the process of growing old.

Alzheimer's disease is the most common dementia in aging. The Alzheimer's dementia syndrome involves cognitive, functional, and behavioral symptoms that most individuals with the disease will exhibit over the course of the illness.

Alzheimer's Symptoms

The following symptoms are characteristic but not all inclusive of Alzheimer's dementia and not always present in each and every case.

Mild symptoms include the following:

1. **memory problems**, not simply forgetting where you left your car keys but forgetting you have a car, losing things, and unable to recall familiar names;
2. **disorientation**, getting lost in the neighborhood, not recognizing familiar places or faces, not dressing correctly, unable to complete simple tasks, loss of abstract thinking, absence of problem-solving ability, planning, and initiative;
3. **problems with routine tasks**, unable to pay bills, avoiding anything new, and forgetting appointments;
4. **mood and personality changes**, ranging from irritability to mild depression, to jealous outbursts and fearful or angry mood swings;

Moderate symptoms include the following:

5. **increased memory loss,** not finding misplaced things, not remembering day of the week, special occasions' names, addresses, favorite places;
6. **poor judgment**, involving clearly inappropriate decisions or actions such as dressing without regard to weather, paying large amounts of money for unneeded products or repairs, and walking through a busy intersection without regard to traffic;
7. **problems with communication**, forgetting simple words or misusing words for wrong objects;
8. **difficulty recognizing family and friends**, forgetting people's name and relationships;
9. **sleep disturbances**, less need for sleep, irregular sleep schedule, and mix-up between night and day;

Severe symptoms include:

10. **loss of appetite, considerable** weight loss, difficulty swallowing, and need of intravenous nutrition;
11. **loss of bowel and bladder control** may need 24/7 care to avoid infections and bedsores;
12. **severe loss of memory**, total dependence on caregiver, loss of mobility, no recognition of people, time or places.

At the severe or final stage, patients require daily and constant assistance as they become immobile and may require support to sit up without falling; many are bedridden. As this stage advances, patients lose the ability to create intelligible speech and a grimace may replace a smile. Body rigidity and loss of range of motion in arms and legs are seen. Sometimes, bladder or bedsore infections can cause a person with dementia to become severely confused, producing delirium, agitation, and hallucinations. Most often, these infections lead to death. Patients can survive the severe stage indefinitely. However, most patients succumb within a year or two from dementia-related causes such as systemic infections, pneumonia, or heart attacks during the course of this stage.

Alzheimer's Diagnosis

Clinicians looking for a possible office diagnosis of Alzheimer's from the symptoms presented by a patient often look for the 4 A's:

Amnesia, loss of memory;
Agnosia, inability to process sensory information such as recognize people, places, or objects;
Apraxia, inability to perform tasks or movements when asked;
Aphasia, inability how to use or understand language.

The 4 A's tend to worsen as dementia progresses and becomes more severe. A medical history, laboratory tests, and a neurological examination are routinely made during the office visit. To rule out or confirm the presumptive Alzheimer's diagnosis made at the medical office, the patient may undergo several more objective tests. These involve neuroimaging of the brain to look for structural or functional damage such as size and atrophy of the hippocampus, white matter lesions, ventricular enlargement, and cortical thinning. Non-enhanced computed tomography (CT) scanning and magnetic resonance imaging (MRI) are appropriate imaging methods that are generally used for detecting reversible causes of mental status changes such as stroke, normal pressure hydrocephalus, brain tumors, and subdural hematomas. Arterial spin labeling functional MRI (ASL-fMRI) has become a preferred noninvasive neuroimaging examination for Alzheimer's because it allows for accurate measurement of cerebral blood flow which is characteristically diminished even before cognitive symptoms appear [14]. While fluorodeoxyglucose positron emission tomography (FDG-PET) scan can be used to evaluate brain function by measuring glucose metabolism (a reflection of synaptic activity), this test is invasive, uses radioactive tracers, and is less accessible than fMRI (see Chap. 15).

Mental status can be assessed using neurocognitive tests. These tests include the Mini-Mental State Examination (MMSE) which provides an evaluation of the cognitive domains affected in Alzheimer's, including recall, attention, orientation;

language and motor planning of new tasks (see Chap. 15). The Mini-Cog test is a brief variation of the MMSE which asks the patient to draw the face of a clock, add the numbers, and insert the clock hands to a specific time indicated by the examiner. The Mini-Cog also asks the patient to remember 3 common objects to recall after a few minutes pause.

There is no single test available to diagnose Alzheimer's although a physician familiar with the disease can determine with a high degree of accuracy whether dementia is present and usually establish whether the dementia is due to Alzheimer's. Neuroimaging tests such as the PiB-PET examination (commercial name Florbetapir) which measures the density of senile plaques in the brain are not recommended for two reasons. First, the finding of senile plaques in the brain is not indicative of Alzheimer's since it has been shown that many cognitively intact individuals have heavy senile plaque formation in their brain. Second, the PiB-PET test accuracy does not increase with time even though cognitive decline may have extensively progressed. Many clinicians do not recommend the PiB-PET test for routine clinical analysis because its diagnostic value is dubious and it is not cost-effective. In addition, false positives that diagnose the presence of Alzheimer's where there is no Alzheimer's can be devastating to one's life. Noninvasive diagnostic tools are reviewed in Chap. 15.

A Caretaker's Story

A vivid and gut-wrenching account of the home care often required to help an Alzheimer's patient finish any small task, such as getting dressed, is given by writer Betty Weiss in her book *Alzheimer's Surgery, An Intimate Portrait* [15]. Her husband, Bernie, had been a NASA engineer and now showed moderate symptoms of Alzheimer's. Weiss writes,

> **In order to be logical, I had to think illogically. Bernie would be sitting and I'd tell him he needed to change his trousers. OK, he'd say and continue sitting. I'd tell him to stand up and he'd say he was standing. It didn't seem possible that someone would not know whether they were standing or sitting, even argue the point, but that's what he'd do. I'd ask if he'd let me help him stand, he'd say 'yes,' then resist my trying to get him up. Trying to get his trousers off while he kept trying to sit down was a real tussle. I'd ask him to undo his belt buckle, and he'd just pull it tighter, I'd pull down the zipper and he'd pull it back up. As the pants started to come down, he'd pull them back up with more strength than I had to pull them down. The more I said 'down' the more he would pull them 'up.' Just as he didn't know the difference between up and down, in and out, off and on, come and go, open and shut. I got to the point where if he said there was a gorilla in the kitchen, I'd reach for a bunch of bananas.** [15]

Weiss's humor, wit, and ability to cope with a most difficult burden likely helped her survive with some degree of sanity the tragic task of being a 24/7 caretaker of a moderately advanced Alzheimer's patient for many years. Many Alzheimer's caregivers are bedeviled with burnout, a stress-related condition involving social withdrawal from friends mixed with anxiety, sleeplessness, exhaustion, and depression. Caregivers are often the victims of ill health as a result of their caregiving duties. Stress associated with caring for a person with dementia, often a 24/7 task, increases a caregiver's chances of developing a chronic illness. When that happens, caregivers are not likely to seek medical help, a fact worsened by poor eating habits, less time for oneself, less time to exercise, and less desire to return to the life they led prior to caretaking.

A Capsule History of Alzheimer's

Alzheimer's disease was first described in a 1907 paper [16] by the German neuropathologist Alois Alzheimer who followed a 51-year-old patient named Auguste Deter (known as Auguste D.), at the Frankfurt Asylum in Germany (Fig. 1.1). According to Alzheimer, this patient showed some symptoms that did not exactly fit any other known mental disorder. After the patient's death in 1906, an autopsy was performed, her brain was examined, and ultrathin sections were put on slides, stained, and examined microscopically. Alzheimer wrote about what he saw.

> **Inside of a cell which appears to be quite normal, one or several fibrils can be distinguished by their unique thickness and capacity for impregnation. Further examination shows many fibrils located next to each other which have been changed in the same way. Next, combined in thick bundles, they appear one by one at the**

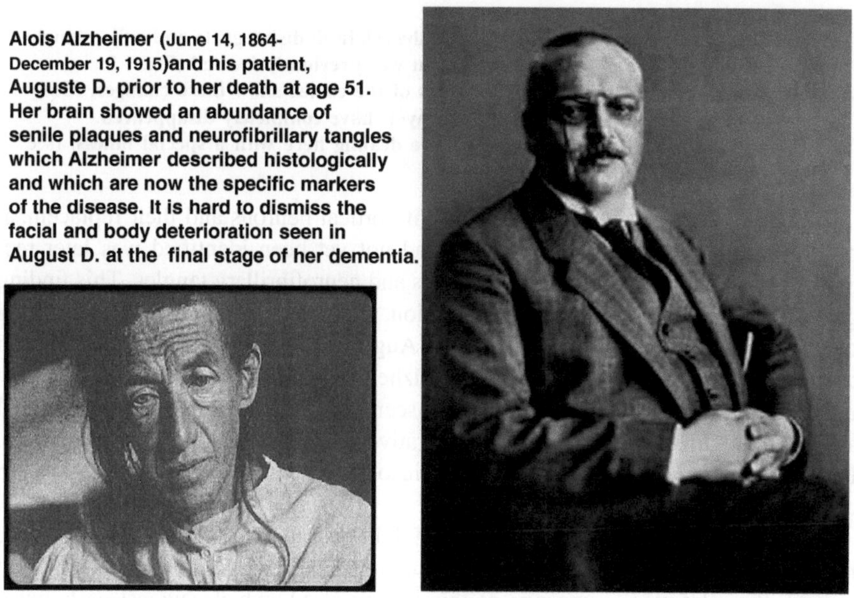

Fig. 1.1 Alois Alzheimer and his patient Auguste Deter

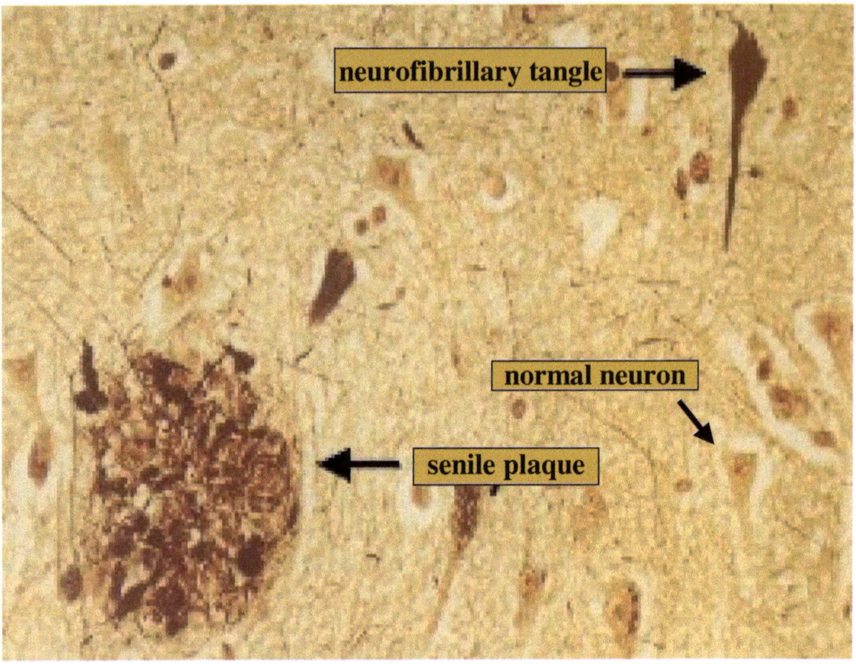

Fig. 1.2 Thin section of Alzheimer's brain tissue showing characteristic senile plaques and neurofibrillary tangles discovered by Alois Alzheimer in 1907 in the brain of Auguste D (see Fig. 1.1)

> **surface of the cell. Finally, the nucleus and the cell itself disintegrate and only a tangle of fibrils indicates the place where a neuron was previously located** (Fig. 1.2).
>
> **Approximately 1/4 to 1/3 of all neurons of the cortex show these changes. Many neurons, especially the ones in the upper layer, have completely disappeared.**
>
> **Considering everything, it seems we are dealing here with a special illness**. (see English translation [17]).

This brief account describing the loss of cortical neurons and their replacement by a group of clumps whose structure had not yet been identified was later recognized as the presence of amyloid plaques and neurofibrillary tangles. This finding gave Alzheimer worldwide name recognition, and he is credited for first describing the unusual pathology seen in the brain of Auguste D (Fig. 1.2). These plaques and tangles are considered the hallmarks of Alzheimer's disease but are not singularly unique to this dementia since they are also seen in cognitively normal aging, as well as in brain trauma and some neurodegenerative disorders such as Down syndrome, Pick's disease, vascular dementia, Parkinson's disease, and Creutzfeldt–Jakob disease.

In looking at Auguste D's picture (Fig. 1.1), one is struck by two details: first, the shriveled face of a person having endured much suffering during her disease and second, the relatively young age of this patient at 51, which suggest early-onset

Alzheimer's not sporadic Alzheimer's, a comparatively rare genetic disorder affecting less than 5 % of all Alzheimer's cases in people mostly under age 50–60. Recently, tissue sections from Auguste D's brain that had been prepared by Alzheimer in 1907 were examined and indicated that Auguste D. had indeed died from early-onset Alzheimer's, but this evidence has been challenged by others [18].

Alois Alzheimer was aware that although many of the symptoms shown by Auguste D. had been previously described in other demented or mentally disabled patients seen at mental asylums and psychiatric hospitals, their brains had not been meticulously examined. The reason for this was probably because staining techniques for brain cells were at their infancy and required patience and perseverance to be of any clinical use. In that regard, Alzheimer was fortunate to have the help of **Franz Nissl**, an experienced and skillful neurohistologist working at the same laboratory that developed a stain for neurons that bears his name and is still used in brain research to this day. Nissl probably introduced Alzheimer to the Bielschowsky silver stain technique that allowed demonstration of senile plaques, neurofibrils, and axons (Fig. 1.2).

In the same year that Alzheimer published his histological observations on the presence of senile plaques and neurofibrillary tangles in the brain of Auguste D, a Czech neuropathologist named **Oskar Fischer** published a clinicopathological study of 16 patients who had died from senile dementia. Fischer described and illustrated in detail the senile plaques in 12 of the 16 patients that characterized senile dementia [19]. His academic affiliation, however, was from the lesser known German University in Prague that was generally in competition with the more influential German University located in Munich and headed by the illustrious **Professor Emil Kraepelin** who credited Alois Alzheimer with the neuropathologic descriptions that bear his name. Had Oskar Fischer worked under Kraepelin as Alois Alzheimer did, the disease would likely be known today as **Fischer's disease** [19].

Just several years before Alzheimer's classic paper describing plaques and tangles, Santiago Ramon y Cajal, the brilliant Spanish neurohistologist, had published his double-impregnation silver stain of brain neurons that partly earned him the Nobel Prize in 1906 for physiology or medicine. For the first time in science, the intricate parts of a neuron were vividly seen, including axons, dendrites, and neurofibrils that stained a prominent black color while other structures of the neuron stained brown on a colorless background (Fig. 1.3). Even the connections neurons made with other neurons were visible using Cajal's silver stain preparations and allowed Cajal to show that the neurons are independent units connected to each other by small contact zones, later called synapses. This was the beginning of our understanding how these nerve cells 'speak' to each other using their branch extensions, called axons and dendrites, in a collaborative effort to interpret information received from outside the brain and processing these signals into meaningful responses.

Since most of the science in the early part of the twentieth century was centered in German universities, Cajal's improved silver stain discovery and his publications in Spanish journals went largely ignored, until his Nobel Prize in 1906.

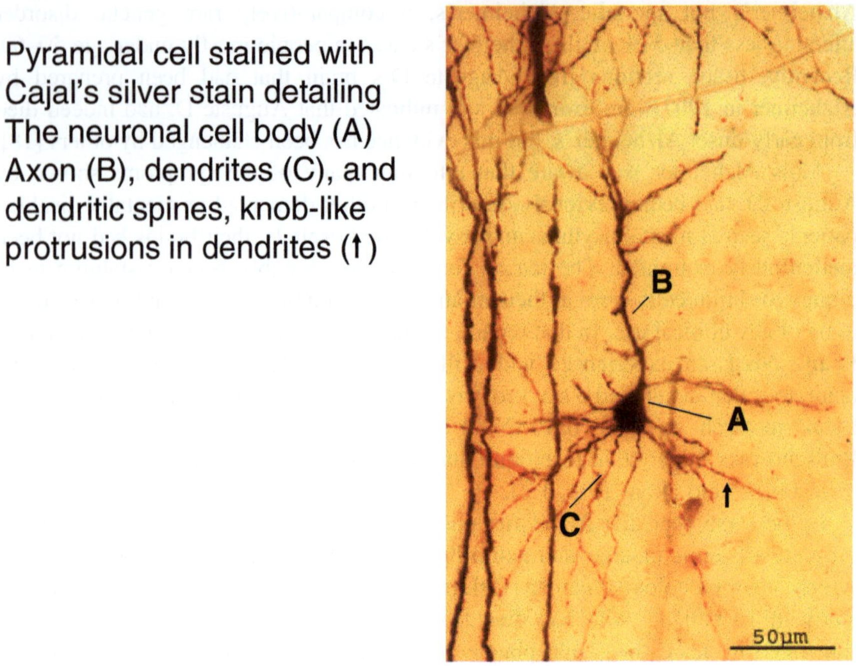

Fig. 1.3 Pyramidal cell stained with Cajal's silver stain detailing. The neuronal cell body (*A*), axon (*B*), dendrites (*C*), and dendritic spines, knob-like protrusions in dendrites (↑)

Nonetheless, very few investigators were aware of Cajal's nerve cell stain in 1906, including Alzheimer. By the same token, it is unlikely that Cajal, not speaking German, was aware of Alzheimer's paper describing dementia. Had that been the case, it is tempting to speculate how much more advancement of the disease may have been made in the years that followed by Cajal and his collaborators.

What Is Dementia?

Dementia is not a disease but a general term that describes a group of symptoms associated with cognitive decline, such as loss of memory, judgment, social abilities and thinking resulting from damage to neurons in the brain. It is therefore a term that should be used only when the specific type of dementia remains undiagnosed or unknown. There are many different types of dementia that have been described. Some common and uncommon ones are listed in Table 1.1.

The most common symptom of all dementias is progressive memory loss. If one considers other symptoms of common dementias, they are almost directly generated by the mild, moderate, and severe forms of dementia. For example, increasing difficulties with performing routine daily tasks, reduced alertness, periods of

Table 1.1 Common and uncommon dementias

Alzheimer's disease (AD)	Most common; progressive degeneration of brain cells characterized by loss of memory and personality changes
Vascular dementia (VaD)	Similar symptoms and pathological changes as AD resulting from small strokes or large brain infarct
Mixed dementia	Similar symptoms as AD and sharing brain pathology from both AD and VaD
Dementia with Lewy Bodies (DLB)	Abnormal protein deposits in brain (Lewy bodies) that disrupt cognitive function and behavior, memory loss less prominent than AD
Creutzfeldt–Jakob Dementia (CJD)	Rare and rapidly progressive viral degeneration of brain cells involving cognitive failure and psychosis caused by misfolded prion protein
Pick's disease (PiD)	Rare, progressive, and slow mental decline caused by Pick's bodies that destroy nerve cells mainly in the frontal lobe (also called frontotemporal dementia)
Wernicke–Korsakoff Syndrome	Mental changes, psychosis and alcoholic encephalopathy caused by vitamin B1 deficiency

confusion, loss of communication skills, impaired judgment, disorientation to time and place, and difficulty with reasoning and problem solving are directly influenced by the degree of memory loss. This pathologic framework can and does introduce behavioral changes in the individual's personality and social interactions including the onset of depression, anxiety, irritability, frustration, emotional outbursts, and physical aggression. These symptoms of dementia can be accelerated by alcoholism, drug abuse, hypertension, abnormal chemical reactions in the body, organ damage from disease, and many other factors. Like most dementias, Alzheimer's structural pathology is characterized by severe brain atrophy, ventricular enlargement, cortical thinning, and extensive neuronal loss in different regions of the brain. These structural changes can flourish over many years and sometimes decades (Fig. 1.4). Structural pathology of the brain is generally associated with metabolic dysfunction.

Metabolic dysfunction of the brain is manifested by neuropsychiatric disturbances which in recent years have been documented by ongoing human clinical studies. These studies have found numerous and complex interactions between metabolic function and the brain. For example, individuals with depression have shown an approximately 60 % higher risk of developing type 2 diabetes. On the other hand, individuals with type 2 diabetes have an elevated risk of developing depression. Metabolic disturbances are also reported to be two to four times higher in people with schizophrenia, and patients prescribed psychotropic medications, such as antipsychotics and antidepressants, often experience disturbances in metabolic-related conditions, including high blood sugar, impaired glucose tolerance, and type 2 diabetes.

Metabolic dysfunction has also been implicated in neurodegenerative disorders, including Alzheimer's, Huntington's, and Parkinson's diseases. It is not known whether metabolic dysfunction precedes or follows neurodegeneration. Multiple

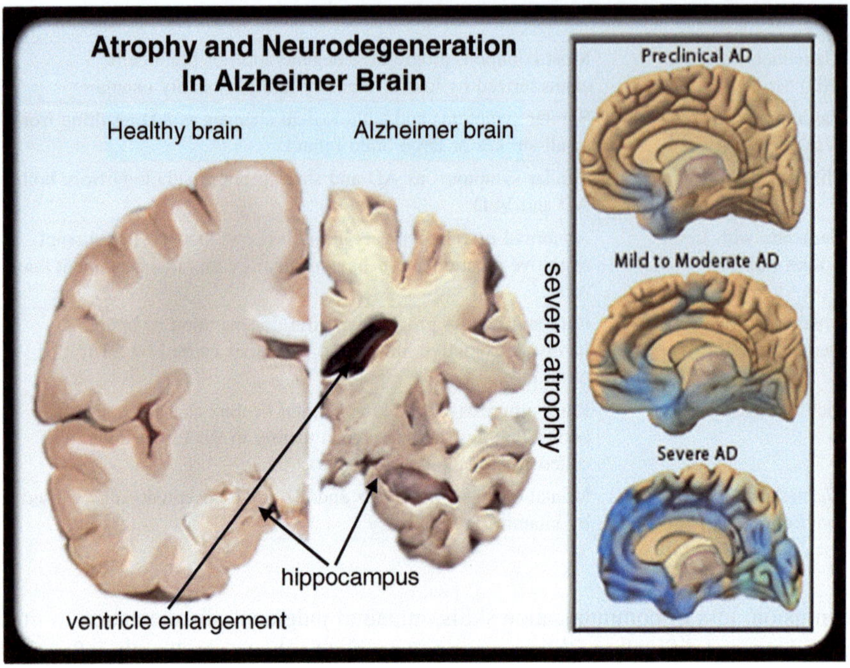

Fig. 1.4 *Right frame* shows where Alzheimer pathology begins (*blue area*) at different stages of the disease. Courtesy of John P. Cunha, DO, FACOEP

clinical observations have demonstrated that dementia in general and Alzheimer's disease in particular are associated with common functional disturbances including sleep apnea, agitated states involving physical or verbal outbursts, general emotional distress, restlessness, and delusions.

Non-Alzheimer's Dementias

Vascular dementia is the second most common dementia after Alzheimer's. Its cause is mainly due to a massive arterial stroke or a series of small vessel strokes (Table 1.1). The word stroke here refers to either blockade of blood flow or bleeding into the brain from a large artery or multiple smaller blood vessels. Brain vessel blockade may be caused by deposits of plaque-like cellular debris that builds up on the inside of the artery wall, or by blood clots that break loose and clog smaller arterioles running deeper in the brain. Vessel bleeding can be due to many factors including very high untreated blood pressure, blood thinner medications, smoking, alcohol abuse, contraceptives, genetic predisposition, and rupture of a brain aneurysm (a ballooning of a vessel by accumulated blood that can stretch the vessel wall until it bursts). This is a rarer form of vascular dementia because half the patients do not survive the hemorrhagic event.

Stroke, due to atherothrombosis of the extracranial carotid arteries, is caused by a combination of factors involving the blood vessels, the clotting system, and hemodynamics. This interaction explains the mechanism of ischemic stroke in patients with carotid atheroma whose outcome is artery-to-artery embolism, low cerebral blood flow, or both.

Diagnosis of vascular dementia is simple at the present time using magnetic resonance imaging (MRI) which can show detailed images of the brain and pinpoint the location of the infarct or vessel bleed from a large artery or smaller microvessels. The test is noninvasive and painless and takes about 20 min to perform.

Neurocognitive and psychometric testing can provide detailed information regarding what mental functions have been affected by the stroke(s). Depending on the location of the vascular damage, these mental changes can affect specific thinking skills such as memory, reasoning, judgment, planning, problem solving, language, and social skills. The vascular dementia signs are quite similar to those seen in Alzheimer's, and we have speculated that both dementias are so similar because they share a common factor, namely reduced blood flow to neurons that are highly sensitive to diminished brain blood flow. It is possible that vascular dementia is a subtype of Alzheimer's disease and that useful treatments aimed at one will benefit the other. These similarities between Alzheimer's and vascular dementia will be explored below.

Vascular Dementia

Table 1.1 reveals some curious similarities between Alzheimer's and vascular dementia that some researchers have described as "mixed dementia" (see next subtopic). As discussed above, vascular dementia is the second (next to Alzheimer's) most common form of dementia known. It develops from blockage or narrowing of blood flow to the brain as a consequence of many small arterioles or a large main artery shutting down. Narrowing or an ischemic stroke of brain vessels can appear as a result of wear and tear from aging or from conditions that favor vessel damage such as diabetes, atherosclerosis, high blood pressure, and some forms of heart disease. It should be noted that a good, non-invasive, easy to perform and accurate test for vascular-related pathology associated with Alzheimer's is available using ankle-brachial index (ABI). This is the main clinical tool to diagnose the presence of lower extremity peripheral artery disease due to atherosclerosis or diabetes. ABI can also be used as an indicator of arterial stiffness caused by hypertension (Fig. 12.7).

Alzheimer's and vascular dementia are extremely difficult to differentiate when a clinical diagnosis is attempted. Both dementias show similar memory loss, risk factors, clinical symptoms, psychological deficits on testing, and interchangeable treatment potential. Both dementias share pathological criteria for either diagnosis [20]. In my judgment, Alzheimer's dementia and vascular dementia should no longer be considered two absolutely different disorders but possible extensions of one dementia overlapping into the other who share a common pathologic process involving brain hypoperfusion [21]. The paradox of the present nomenclature that classifies Alzheimer's and vascular dementia as different disorders has to do with their supposed causal origin. On one side of the issue, many investigators classify Alzheimer's as a neurodegenerative disease which later threatens damage to the cerebrovascular system. This group also believes that vascular dementia is not a neurodegenerative disorder but has a vascular component that later threatens or accelerates neurodegeneration.

A second group of investigators has proposed a solution to this dilemma by classifying a patient with both plaques and tangles and vascular pathology in the brain as a 'mixed dementia'. Although it would appear that 'mixed' dementia is more common than either dementia, this comorbid state does not explain how Alzheimer's and vascular dementia are expressed into one entity (mixed) unless a common causal factor unites them. Commonalities aside, the clinical diagnosis of Alzheimer's and vascular dementia based on clinical symptoms alone is at best, very difficult to make, according to many specialists with expertise in these disorders.

A third group of investigators, present author included, believe that both Alzheimer's and vascular dementia are caused by poor blood flow available to the brain and are therefore vascular disorders that are prone to provoke neurodegenerative pathology. This disagreement among investigators is not merely an exercise in tautological nonsense because professional reputations, grant funding, and pharmaceutical pursuits of drugs to combat neurodegenerative changes in Alzheimer's disease are at stake. Again, it is a question of money and vanity in preference to evidence. In Chap. 15, this issue is examined in detail and evidence-based findings derived from a horde of studies will be presented that favor the conclusion that Alzheimer's and vascular dementia are initiated by vascular factors.

This problem of correctly classifying both dementias according to probable cause continues to fester not only because diagnostic criteria of dubious reliability are ever present but also because overlapping or identical pathology is found in both disorders. For example, Alzheimer's and vascular dementia share features involving brain hypoperfusion, white matter lesions, and pathological markers typical in both disorders and genetic links [20]. The use of computer tomograms (CT) or magnetic resonance imaging (MRI) brain scans contributes little to characterizing either dementia when lesions in the white matter of the brain are present or when brain atrophy is involved. Additionally, when brain vessel abnormalities are present, less Alzheimer's pathology is seen in postmortem examinations [22], a finding that indicates cerebral ischemia is a critical factor in the development of both Alzheimer's and vascular dementia. Not even postmortem examination of Alzheimer's and vascular dementia brains is able to clearly provide a final diagnosis between these two afflictions. The reason is that at present, necropsy identification of Alzheimer's contains a common fallacy; it includes cases in which Alzheimer's disease exists with other diseases affecting cognition, including vascular dementia [23].

The distinction or similarities between Alzheimer's and vascular dementia were one of the early clues in our thinking that both dementias shared a common and fundamental factor: poor blood flow in the brain. Although this concept has become highly contentious, it must provide some pause when common therapy is designed that can prevent or alleviate either disorder.

Mixed Dementia

Mixed dementia is a relatively new term that describes cognitive deficits similar to both Alzheimer's and vascular dementia. This similarity also applies to the neurodegenerative lesions seen in both dementias. The degenerative pathology contained in mixed dementia **is a combination of brain tissue abnormalities** that characterize both disorders, that is, **plaques and tangles, loss of neuronal synapses, and cerebrovascular disease.** Some investigators believe that mixed dementia is more common than either Alzheimer's or vascular dementia [24]. Other investigators deny that mixed dementia even exists. It is possible that vascular dementia developing from small strokes is a variant of Alzheimer's disease that acts as an accelerant of cognitive meltdown in a much more rapid manner than the chronic brain hypoperfusion that precedes most dementias, including Alzheimer's.

Our opinion is that mixed dementia is the most characteristic version of Alzheimer's disease since it is extremely unlikely that 'pure' Alzheimer's exists, that is, a brain with plaques and tangles but no visible vascular pathology. Vascular-related changes in Alzheimer's brains are not entirely due to vascular dementia which is usually symptomatic. Vascular brain pathology affecting Alzheimer's is often asymptomatic, containing white matter lesions, silent strokes, atherosclerotic vessels, and cerebral amyloid angiopathy (plaque-like material inside vessel walls).

These asymptomatic but damaging conditions to brain tissue are highly prevalent in Alzheimer's and mixed dementia. If a pathologist carefully examines an Alzheimer's brain after death and observes no signs of strokes, no white matter lesions, no vessel atherosclerosis, and no amyloid angiopathy, there could still be present an important vascular contribution called **chronic brain hypoperfusion** that cannot be discerned by gross examination. This 'unobservable factor' could lead to the erroneous conclusion that 'pure' Alzheimer's pathology is present and falsely inflates epidemiological findings. The postmortem examination of whether chronic cerebral hypoperfusion is present or not can be corrected by measuring cerebral blood flow at some point prior to death, although this is not always possible. Certain brain regions in elderly people who have not yet expressed Alzheimer's symptoms have shown markedly reduced brain blood flow when measured with MRI techniques or Doppler ultrasound [14, 25, 26]. Chronic cerebral hypoperfusion, therefore, represents an important marker of impending cognitive decline and possibly the major cause of Alzheimer's and so-called mixed dementia.

More than 2 dozen conditions are known to induce chronic brain hypoperfusion, otherwise known as vascular risk factors for Alzheimer's disease. These conditions include diabetes type 2, peripheral artery disease, carotid atherosclerosis, high blood pressure, high serum cholesterol, high serum lipids, sleep apnea, and a variety of heart disease conditions [27]. Without exception, each of these conditions contributes to diminishing blood flow to the brain which in the elderly individual can spell cognitive dysfunction. Physicians should be aware that measuring cerebral blood flow when encountering any of these vascular risk factors should be performed in case treatment needs to be applied or possibly modified. Blood flow to the brain is easily measured with magnetic resonance imaging (MRI) by quantifying the volume of blood passing through the microvascular network in a given volume of tissue over a certain period of time.

Cerebral Blood Flow Diminishes with Age

As a person normally ages, cerebral blood flow diminishes by a half percent (0.05 %) every year so from age 22–62, blood flow to the brain has dropped 20 % [28].

When any of the disorders named above are contracted by an elderly individual, another 10–15 % reduction in cerebral blood flow can result in an additional burden to the delivery of blood flow to brain cells. This situation jeopardizes brain cell survival because oxygen and glucose from blood flow delivery are crucial to maintain normal brain activities, including cognitive function. When brain blood flow falls below a critical level in an elderly person, the supply and demand homeostasis (called 'neurovascular coupling') established by cerebral blood flow and brain cell metabolism will be upset. The 'uncoupling' of blood flow metabolism will eventually bring dire consequences to brain function including neurodegenerative changes, brain cell loss, cognitive decline, and brain atrophy, all merciless contributors of dementia.

Mixed dementia as a subtype of Alzheimer's disease clearly supports the far-reaching conclusion that Alzheimer's disease is primarily a vascular disorder that with time induces neurodegeneration and cognitive deterioration. If Alzheimer's is to be considered primarily a vascular disorder, a link between its preclinical detection, its risk factors, its prodromal features, and its treatment opportunities requires a strong association with cerebral perfusion insufficiency. These requirements are met according to multidisciplinary studies that support the vasculopathic link between the cognitive impairment that precedes Alzheimer's and brain hypoperfusion [29]. For this reason, the term '*vasocognopathy*' was coined in one of my papers [21] which serves to describe that Alzheimer's begins as a vascular disturbance (vaso-) affecting cognitive function (cogno-) in a pathological way (pathy). Brain hypoperfusion has been shown to induce the formation of Abeta in cortical neurons and axons in mice models engineered genetically to mimic Alzheimer's neuropathology [30]. This and similar experimental findings indicate that chronic brain hypoperfusion precedes and likely generates Abeta formation in Alzheimer's brain. Amyloid deposition in the brain of older people is also enhanced by vascular risk factors [31] Uncontrolled hypertension, which is a major vascular risk factor for Alzheimer's and for cerebral hypoperfusion, is reported to boost Abeta deposition in the cortex when compared to non-hypertensive controls [31].

Dementia with Lewy Bodies

About 15–25 % of demented patients are found at autopsy to have Lewy bodies in the brain. These inclusion bodies are found composed of densely packed filaments and granules which insert inside neurons of the cortex and are mainly seen in elderly patients with Parkinson's disease. Lewy body dementia resembles Alzheimer's symptoms, but the inclusions are neither plaques nor tangles but made up of a protein called synuclein. Lewy body accumulations are speculated to contribute to cognitive impairment in elderly demented individuals, but they can also be present in the brains of cognitively intact persons.

This fact has created some controversy as to the actual role of Lewy bodies in dementia and the reason is partly because their structure and composition remain basically unknown. Curiously, many Lewy bodies are found in the basal nucleus of Meynert (see **cholinergic hypothesis**) which is a brain region targeted by Alzheimer's neurodegenerative changes that result in the death of many cholinergic neurons and their connections to other brain regions. Destruction of the basal nucleus of Meynert by Lewy bodies tends to promote Alzheimer's symptoms. When Lewy bodies are found in areas of the brain that contain dopamine neurons, Parkinson's disease is likely to develop. Lewy body dementia symptoms seem to progress more rapidly than Alzheimer's symptoms and unlike Alzheimer's, can involve movement problems, visual hallucinations, and blood pressure changes.

As with Alzheimer's, dementia with Lewy bodies has no cure. Treatments are available for symptomatic relief of psychiatric and motor symptoms. Psychiatric symptoms include hallucinations, depression, and delusions and sleep disorders. Other symptoms such as excessive daytime drowsiness, repeated falls and syncope, and loss of smell are common. Medications to control these symptoms often produce adverse reactions, and minimal doses prescribed by the attending physician that will control or manage the symptoms are essential to prolong good quality life.

Creutzfeldt–Jakob Dementia

Creutzfeldt–Jakob dementia is a rare infectious disorder caused by a protein called prion. Prions are usually harmless in the body but when they undergo structural changes, they can cause a number of disorders including 'mad cow disease,' as well as an infectious disease affecting goats and sheep called scrapie. Creutzfeldt–Jakob dementia also is seen in kuru, which is found among natives of New Guinea who practiced a form of cannibalism in which they ate the brains of dead people with infectious prion as part of a funeral ritual. That practice was outlawed in the early 1960s.

Consequently, infectious prion can be transmitted from contaminated food, blood transfusions, and medical procedures involving surgery of contaminated tissue [32]. Creutzfeldt–Jakob dementia can be acquired from inheritance of a parent with the disorder and more commonly from exposure to infectious prion but not from casual contact. When Creutzfeldt–Jakob dementia is acquired, usually after a long asymptomatic incubation period that can range from several years to 50 years, it seems to come out of the blue with no single agent appearing as the trigger. It ravages the brain causing holes in the tissue much like Swiss cheese, called spongiform encephalopathy [32]. Rapid changes in personality and physical disabilities appear as memory, speech, reasoning, thinking, and other cognitive skills disintegrate. Death follows from complications involving heart failure, respiratory insufficiency, or pneumonia ending life within 12 months. Treatment is fundamentally useless since it only extends the misery of the disease by a few months. Drugs for pain and seizures are available, but their effect is mildly palliative.

Frontotemporal Dementia (Pick's Disease)

This is a dementia that mainly affects the frontal and temporal lobes of the brain which does not afflict the elderly as Alzheimer's dementia does. Frontotemporal dementia or Pick's disease can occur in people as young as 20 but is more common between age 40 and 60. The dementing symptoms appear slowly and initially involve personality changes of the individual. This is a clue that Alzheimer's is not a likely diagnosis since memory impairment in Pick's comes much later. Initially, Pick's disease is characterized by language problems and deteriorating social skills. Later in the evolution of Pick's disease, memory loss, lack of concentration, emotional dullness, and difficulty thinking dominate the clinical picture. Pick's bodies are found distributed inside neurons that contain an abnormal form of a protein called tau. Tau is actually a normal protein found in all nerve cells whose function is to assemble the 'railroad tracts' or microtubules, hollow shafts within axons and dendrites that transport to and from the cytoplasm, and the nerve ending material necessary to keep the neuron alive. But an excessive amount of tau can form to destroy the cell by a process not yet known. When tau mutation occurs in one parent, the disease can be passed on to the progeny as an autosomal dominant.

Wernicke–Korsakoff Syndrome

The dementia symptoms caused by Wernicke–Korsakoff syndrome are truly reversible if caught in time. Alcoholic abuse is generally one of the culprits which can deplete vitamin B1 (thiamine) stores to a level that induces disorientation to time and place, confusion, visual problems, and profound loss of memory and inability to form new memories. If extensive irreversible brain damage has not yet occurred, oral thiamine and vitamin B complex should be given indefinitely. Intravenous or intramuscular thiamine can be given in a primary care setting. When alcoholism is not found to be the cause, gastrointestinal tests and nutritional intake should be evaluated to confirm a thiamine deficiency from other potential disorders that can mimic this syndrome.

References

1. Alzheimer's Association. 2014 Alzheimer's disease facts and figures. Alzheimer's Dement. 2014;10(2):16–20.
2. de la Torre JC, Mussivand T. Can disturbed brain microcirculation cause Alzheimer' disease? Neurol Res. 1993;15:146–53.
3. de la Torre JC. The vascular hypothesis of Alzheimer's disease: bench to bedside and beyond. Neurodegener Dis. 2010;7(1–3):116–21.
4. Davis DG, Schmitt FA, Wekstein DR, Markesbery WR. Alzheimer neuropathologic alterations in aged cognitively normal subjects. J Neuropathol Exp Neurol. 1999;58:376–88.
5. Bryan KJ, Lee H, Perry G, Smith MA, Casadesus G. Transgenic mouse models of Alzheimer's disease: behavioral testing and considerations. In: Buccafusco JJ, editor. Methods of behavior analysis in neuroscience. 2. Boca Raton (FL): CRC Press; 2009.
6. Robakis NK. Are Abeta and its derivatives causative agents or innocent bystanders in AD? Neurodeg Dis. 2010;7(1–3):32–7.
7. Lee HG, Castellani RJ, Zhu X, Perry G, Smith MA. Amyloid-beta in Alzheimer's disease: the horse or the cart? Pathogenic or protective? Int J Exp Pathol. 2005;86(3):133–8.
8. Morris GP, Clark IA, Vissel B. Inconsistencies and controversies surrounding the amyloid hypothesis of Alzheimer's disease. Acta Neuropathol Commun. 2014;18(2):135.
9. Snitz BE, Weissfeld LA, Lopez OL, Kuller LH, Saxton J, Singhabahu DM, Klunk WE, Mathis CA, Price JC, Ives DG, Cohen AD, McDade E, Dekosky ST. Cognitive trajectories associated with beta-amyloid deposition in the oldest-old without dementia. Neurology. 2013;80:1378–84.
10. Rentz DM, Locascio JJ, Becker JA, Moran EK, Eng E, Buckner RL, Sperling RA, Johnson KA. Cognition, reserve, and amyloid deposition in normal aging. Ann Neurol. 2010;67:353–64.
11. Hedden T, Oh H, Younger AP, Patel TA. Meta-analysis of amyloid-cognition relations in cognitively normal older adults. Neurology. 2013;80:1341–1348. (plaques unrelated to AD).
12. Engler H, Forsberg A, Almkvist O, Blomquist G, Larsson E, Savitcheva I, Wall A, Ringheim A, Långström B, Nordberg A. Two-year follow-up of amyloid deposition in patients with Alzheimer's disease. Brain. 2006;129:2856–66.
13. Ferri CP, Prince M, Brayne C, Brodaty H, Fratiglioni L, Ganguli M, et al. Global prevalence of dementia: a Delphi consensus study. Lancet. 2005;366:2112–7.

14. Xekardaki A, Kövari E, Gold G, Papadimitropoulou A, Giacobini E, Herrmann F, Giannakopoulos P, Bouras C. Neuropathological changes in aging brain. Adv Exp Med Biol. 2015;821:11–7.
15. Weiss B. Alzheimer's surgery: an intimate portrait. Bloomington, Indiana: AuthorHouse; 2005.
16. Alzheimer A. Uber eine eigenartige Erkrankung der Hirnrinde. Allgemeine Zeitschrift für Psychiatrie und phychish-gerichtliche Medizin, (Berlin) 1907;64:146–48.
17. Bick K, Amaducci L, Pepeu G, editors. Translation of the historic papers by Alois Alzheimer, Oskar Fischer, Francesco Bonfiglio, Emil Kraepelin, Gaetano Perusini. New York: Raven Press; 1987. The early story of Alzheimer's disease.
18. Muller U, Winter P, Graeber M. A presenilin 1 mutation is the first case of Alzheimer's disease. Lancet Neurol. 2013;12:128–9.
19. Goedert M. Oskar Fischer and the study of dementia. Brain. 2009;32:1102–11.
20. de la Torre JC. Alzheimer's disease: how does it start?J Alzheimers Dis. 2002;4(6):497–512.
21. de la Torre JC. Alzheimer's disease is a vasocognopathy: a new term to describe its nature. Neurol Res. 2004;26(5):517–24.
22. Snowdon DA, Greiner LH, Mortimer JA, Riley KP, Greiner PA, Markesbery WR. Brain infarction and the clinical expression of Alzheimer disease. Nun Study JAMA. 1997;277 (10):813–7.
23. Bowler JV, Eliasziw M, Steenhuis R, Munoz DG, Fry R, Merskey H, Hachinski VC. Comparative evolution of Alzheimer disease, vascular dementia, and mixed dementia. Arch Neurol. 1997;54(6):697–703.
24. Korczyn AD. Mixed dementia—the most common cause of dementia. Ann N Y Acad Sci. 2002;977:129–34.
25. Ruitenberg A, den Heijer T, Bakker SL, van Swieten JC, Koudstaal PJ, Hofman A, Breteler MM. Cerebral hypoperfusion and clinical onset of dementia: The Rotterdam Study. Ann Neurol. 2005;57:789–94.
26. Alsop DC, Dai W, Grossman M, Detre JA. Arterial spin labeling blood flow MRI: its role in the early characterization of Alzheimer's disease. J Alzheimers Dis. 2010;20(3):871–80.
27. de la Torre JC. Is Alzheimer's disease a neurodegenerative or a vascular disorder? Data, dogma, and dialectics. Lancet Neurol. 2004;3(3):184–90.
28. Leenders KL, Perani D, Lammertsma AA, Heather JD, Buckingham P, Healy MJR, Gibbs JM, Wise RJS, Hatazawa J, Herold S, Beaney RP, Brooks DJ, Spinks T, Rhodes C, Frackowiak RS, Jones T. Cerebral blood flow, blood volume and oxygen utilization. Brain. 1990;113:24–47.
29. Viticchi G, Falsetti L, Buratti L, Boria C, Luzzi S, Bartolini M, Provinciali L, Silvestrini M. Framingham risk score can predict cognitive decline progression in Alzheimer's disease. Neurobiol Aging. 2015;36(11):2940–5.
30. Kitaguchi H, Tomimoto H, Ihara M, Shibata M, Uemura K, Kalaria RN, Kihara T, Asada-Utsugi M, Kinoshita A, Takahashi R. Chronic cerebral hypoperfusion accelerates amyloid beta deposition in APPSwInd transgenic mice. Brain Res. 2009;19(1294):202–10.
31. Rodrigue KM. Contribution of cerebral health to the diagnosis of Alzheimer's disease. JAMA Neurol. 2013;70:438–9.
32. Liberski PP, Sikorska B, Hauw JJ, Kopp N. Ultrastructural characteristics (or evaluation) of Creutzfeldt-Jakob disease and other human transmissible spongiform encephalopathies or prion diseases. Ultrastruct Pathol. 2010;34:351–61.

Chapter 2
Storing and Keeping Memories

Forming Memories

How memories are formed in the brain is an extremely complex process that is only partly understood. Memories are fundamental for learning and being able to interact with our environment. There is a sequence of events that are involved in forming memories. These events include the acquisition and storage of something one wishes to remember and after retaining this information, being able to retrieve it when there is a need to use it. There are two types of memories that the brain uses to store and retrieve a memory for brief or long durations. The first type is **short-term memory**. Short-term memory, as the name implies, is holding a small amount of information that is acquired (typical about 7 items or less), briefly storing it and then usually forgetting it after about 30 s. Short-term memory can be retained by repetition that is how one is able to retain a telephone number that is important to recall. Short-term memory information can be transferred to long-term memory within seconds, but the mechanisms involved in this transformation remain controversial.

The second type of memory is called long-term. Long-term memory is acquired and stored permanently. The storage and retrieval of long-term memories can last a few days too many years. How well one remembers memories that are heard, seen, smelled, felt or tasted depends on how precise the brain can record them. Weak memories only come to mind when prompting or reminding.

Procedural, episodic, and semantic memory is part of long-term memory.

Procedural memory is involved in knowing how to do things, such as riding a bicycle or putting your shoes on. It does not require a conscious component and can be performed without thinking.

Episodic memory is a part of the long-term memory responsible for acquiring and storing events (episodes) that have been experienced in life, such as cutting your hand with a knife or the first time you wore eyeglasses. Most adults in the nation recall what they were doing on 9/11 when the xxx Towers went down.

J.C. de la Torre, *Alzheimer's Turning Point*,
DOI 10.1007/978-3-319-34057-9_2

Semantic memory also participates in long-term memory and is the ability to store information that includes general knowledge and the meaning of words.

Episodic and semantic memories but not procedural memory are declarative or explicit, that is, they require conscious thought, for example, recalling facts and events such as Napoleon's war in Russia or solving quadratic equations. Procedural memories are implicit, they include memories for skills and habit. Pavlovian conditioning is an example of procedural memory where a dog learns to salivate at the ringing of a bell that signifies food is coming.

Creating a Memory

A memory perceived by the senses is said to be **encoded**, a process that constitutes the first step in creating and storing a memory. Encoding takes place in various areas of the frontal cortex and other cortical regions of the brain where they are decoded (Fig. 2.1).

Following decoding, the memory travels to a part of the brain called the **hippocampus**, a structure deep within the medial temporal lobe of the brain responsible for **learning and memory**. Learning is described as the process by which new information about the environment is acquired and memory as the mechanism by which that knowledge is retained. As memories arrive at the hippocampus, a decision is made as to whether or not the memory is worthy of being permanently stored in the cortex for potential recall. It is generally believed that the cortex is capable of supporting various types of memories acquired from visual, hearing, olfactory, tactile, and auditory stimuli. This notion is supported by many studies using functional neuroimaging techniques that light up the screen following neuron activation in different parts of the cortex when the brain is exposed to various sensory or mental stimuli.

For long-term storage, memories undergo a process called **consolidation**. Memory consolidation is the means by which a memory is stabilized after its initial acquisition. Following consolidation of a memory, **long-term potentiation** (LTP) may take place. LTP is believed to prolong the strengthening mechanism of synaptic transmission and to maintain memories in the brain for long durations (Fig. 2.1).

This process involved in memory is highly complex and incompletely understood. Memory storage and retrieval involves thousands of neural connections, neurotransmitter release, protein synthesis and the participation of a variety of chemical molecules. The role of the hippocampus in humans was studied in Henry Molaison (known in neuroscience as "H.M."). Molaison was a man, who due to intractable seizures, had his hippocampus surgically removed bilaterally. After the operation, he was seizure free, but it was noted that he had great difficulty in forming new memories, learning new words, and remembering what had happened a short time before. However, older memories of facts and events formed before his operation could be recalled, a detail that suggested storage of long-term memories

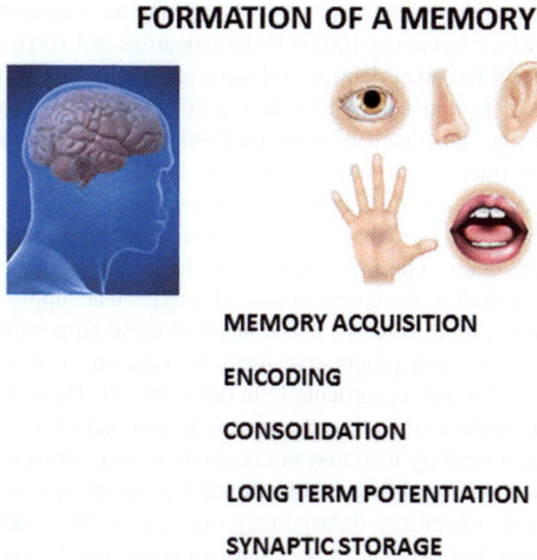

Fig. 2.1 Memories are acquired from sensory stimuli derived from the 5 senses, seeing, smelling, hearing, feeling, and tasting. The first step in creating a memory involves encoding, a process where the perceived item of interest is converted into a construct that can be stored in the brain. After encoding, the memory trace is decoded in different regions of the sensory cortex where it is eventually sent to the hippocampus, a small region of the brain that elects whether the memory will be stored long-term or discarded. Long-term storage of the memory involves consolidation, where the memory is stabilized and stored using a process called long-term potentiation. Retrieval or recall of a memory in long-term storage is based on cues that prompt the act of remembering. Short-term memory undergoes a similar process as long-term memory but differs essentially due to its brief duration and limited capacity of items to be stored. Short-term memory quickly decays within 30 s unless a conscious effort is made to retain the information

did not reside in the hippocampus. Where in the brain is memory stored has been a topic of much research and considerable debate. Evidence indicates that long-term memory is processed in the prefrontal cortex and then flows to different regions of the brain by neural circuits and networks ending up in the synapse, the connecting point between neurons.

This long-held belief has been challenged recently by **Dr. David Glanzman**, a professor of Integrative Biology and Physiology at UCLA whose work with the 5-inch long Aplysia snail, suggests that long-term memory is not stored in synapses. Glanzman's research has reported that although synapses participate in long-term memory that is not where memories are stored [1]. Glanzman suspects that memories instead may be stored in the nucleus of neurons, although this has not been proven [1]. Knowing where long-term memory is stored is of key importance in research because it can provide targets for restorative therapy in neurons or synapses that have not been killed by trauma or neurodegeneration. That is the advantage of using a simple organism like Aplysia which have only about

20,000 neurons when doing experiment involving long-term memory compared to the human brain which has about 100,000 billion neurons and 100 trillion synapses.

How Alzheimer's and other dementias destroy short- and long-term memory has been the subject of heated controversy in recent years. It may be assumed that hippocampal damage and death of neurons from the neurodegenerative process involved in Alzheimer's and other dementias is an important contributor of short-term memory loss and eventually of long-term memory.

Rodent experiments in our laboratory have shown that hippocampal neurons are the first cells in the brain to perish several weeks following reduced cerebral blood flow induced by occluding the common carotid arteries that supply about 60 % of the blood flow to the brain (Fig. 4.1). The death of these hippocampal neurons in rodents was associated with progressive memory impairment, reduced ATP, and eventual atrophy of the cortex, particularly in older rats [2]. The delicate sensitivity of hippocampal neurons to mild, but chronic brain hypoperfusion remains unclear.

It is known that a memory item may not be retained once damage to neurons that encode or decode a memory from sensory stimuli has occurred. It follows that tests involving the degree of sensory damage may offer a clue that a dementia process may be forthcoming. For example, a poor sense of smell may be one of the earliest signs of Alzheimer's, especially when it correlates with other mental and cognitive deficits characteristic of this dementia.

For now, it is intriguing to know that memory loss can occur not only from dementia but also from physical, physiological, and psychological trauma to the brain. These very different memory-robbing mechanisms imply that a disturbed common element may explain the damage to neurons, their connections, and their function. That common element may involve inadequate blood flow to the brain, among other things. I will present this argument throughout the balance of this book.

References

1. Cai D, Pearce K, Chen S, Glanzman DL. Reconsolidation of long-term memory in Aplysia. Curr Biol. 2012;22(19):1783–8.
2. de la Torre JC, Fortin T, Park GA, Saunders JK, Kozlowski P, Butler K, de Socarraz H, Pappas B, Richard M. Aged but not young rats develop metabolic, memory deficits after chronic brain ischaemia. Neurol Res. 1992;14(2 Suppl):177–80.

Chapter 3
Masquerading as Dementia

When it is not Dementia

Distinguishing one type of dementia from another may be difficult without running many tests. However, **distinguishing dementia from delirium** or some other non-dementia condition presenting similar symptoms as those seen in dementia is crucial since such symptoms may be reversible after treatment. Delirium, for example, can usually be traced to a contributing factor such as medications, brain trauma, alcohol or drug abuse, surgery, infection, and a host of medical illnesses associated with high fever. Fever can cause delirium because high temperature (105° F and over) can interfere with multiple processes involving brain metabolism.

Delirium is generally accompanied by confusion, disorientation, memory difficulties, an inability to focus one's attention, and hallucinatory episodes. Delirium can often be quickly differentiated from dementia because it appears suddenly and the person cannot be easily aroused to an alert state. A physician can diagnose delirium with the help of a close relative who can describe the sudden beginnings of this condition.

Other conditions that mimic dementia are vitamin B12 deficiency. When vitamin B12 levels are chronically low, memory impairment, agitation, and depression can be easily confused with early signs of Alzheimer's. This error can lead to the patient being prescribed Alzheimer's medication which can complicate or worsen the clinical symptoms. An easy blood test for vitamin B12 can avoid this misdiagnosis, and treatment with daily vitamin B12 supplement generally reverses the mental impairment.

J.C. de la Torre, *Alzheimer's Turning Point*,
DOI 10.1007/978-3-319-34057-9_3

Normal Pressure Hydrocephalus

Normal pressure hydrocephalus (NPH) is a brain disorder where an abnormal buildup of cerebrospinal fluid (CSF) collects in the brain's ventricles or cavities. This buildup within the ventricles occurs because normal CSF absorption is blocked in some way. This causes the ventricles to enlarge and exert pressure on the brain. As brain pressure increases, mental sharpness decreases, resulting in thinking difficulties, impaired decision-making, reduced concentration, and personality changes. These symptoms mimic Alzheimer's disease [1]. NPH can occur at any age but is more commonly seen in the elderly. Early diagnosis by a neurologist with expertise in mental disorders improves the chances for good recovery. A history of conditions that can damage the brain such as hemorrhage, infection, tumors, and trauma are often precursors of NPH.

If not treated in time, NPH can cause brain damage and death. Surgical treatment is generally applied using a thin catheter (shunt) inserted into the ventricle to drain excess CSF into the abdomen. Good recovery of mental problems can be seen within days following shunt surgery, but in some patients, improvement is not seen after weeks, months, or ever. Even after good recovery from NPH, some patients may decline again requiring a revision of the original shunt surgery.

Thyroid Disease

An underactive thyroid gland can result in symptoms that resemble dementia. The thyroid is a butterfly-shaped gland that sits in front of the windpipe in the neck and produces two hormones, T3 and T4. These hormones regulate basal metabolism in the body whose main function is to break down food and convert it to energy. When T3 and T4 are underproduced**, hypothyroidism** can result leading to psychiatric symptoms involving diminished cognition and memory. Other mental symptoms of hypothyroidism are fatigue, disturbed sleep, and depression which in an elderly individual can easily lead to a diagnosis of dementia. When hypothyroidism is diagnosed, treatment to control thyroid deficiency usually reverses any mental impairment. Replacement therapy with oral T4 remains the treatment of choice. When patients' cognitive dysfunction does not rapidly improve with T4, T3 may be added.

When **hyperthyroidism** is diagnosed, a spectrum of psychological disturbances can appear that resemble dementia symptoms. These mental disturbances are similar to the state of hypothyroidism and include depression, cognitive deterioration, emotional mood swings, inability to concentrate on problems, and a hypomanic state which is temporary episodes of mild mood swings uncharacteristic of the person but not sufficiently severe so as to cause social or work-related problems.

Panic attacks are common in hyperthyroid disease and can precede hyperthyroid activity by many years. The most common causes of hyperthyroidism are Graves' disease goiter and thyroid nodules. Graves' disease involves malfunction of the immune system where thyroid hyperactivity produces a pounding fast heart rate, profuse sweating, and weight loss. When properly treated, Graves' disease symptoms disappear after a few months or years. An enlargement of the thyroid called goiter appears when there is a lack of iodine in the diet or more commonly in the US, high or low production of thyroid hormones. Thyroid nodules can arise as a single lump or multiple small lumps in the thyroid to cause overproduction of thyroid hormones and create symptoms of hyperthyroidism. Treatment may include periodic observation, drug therapy to block hormone production, or radioactive iodine that destroys thyroid tissue. Surgery to remove the thyroid gland is another medical option. These treatments generally lead to hypothyroidism which can be treated with a synthetic hormone tablet called thyroxine. The psychiatric symptoms of hyperthyroidism can be substantially diminished or reversed when the rate of thyroid hormone production is therapeutically controlled [2].

Alcoholism

Alcoholism is a great imitator of dementia because it can worsen or mimic practically any psychiatric condition seen in the mental health setting. The likely reason for this is that heavy alcohol use directly affects brain function and alters various brain chemicals, namely neurotransmitters and hormonal systems known to be involved in the development of many common mental disorders.

Neurotransmitters are chemical messengers that transmit signals throughout the body that control thought processes, behavior, and emotion. An important neurotransmitter altered by alcohol is dopamine. Some investigators believe dopamine acts on the brain like natural cocaine, affecting cognition that can lead to euphoric and emotional states.

Alcohol abuse increases dopamine levels by binding to dopamine receptors in neurons producing feelings of pleasure. When too much dopamine is produced, euphoria, a hypomanic state, confusion, and psychosis can result. This latter condition is known as alcohol psychosis, and it manifests itself by the appearance of hallucinations, paranoia, and delusions. Chronic abuse of alcohol can lead to Wernicke–Korsakoff syndrome, a condition due in part to thiamine deficiency (vitamin B1) that is expressed by the patient confabulating psychosocial situations that do not exist while also affecting learning and memory loss. Alcohol-related psychosis often clears when alcohol use is discontinued but can return with a vengeance when alcoholism is resumed [3].

There are more than 50 conditions and disorders that can mimic some or most dementia symptoms but the majority of these dementia "imitators" can be easily diagnosed and treated successfully by an experienced neurologist or specialist in neurodegenerative-related conditions. When that happens, the dementia-like symptoms are generally reversed. These conditions that mimic dementia can involve nutritional deficiencies, infections, trauma, and some forms of cancers.

Treating Alzheimer's

There are four FDA approved oral medications currently used for treating Alzheimer's patients. In their best light, they can modestly manage Alzheimer's symptoms to a point, but they will not alter the course of the dementia nor reduce its severity. The 4 drugs differ only slightly from an inert medication.

These are 3 cholinergic drugs that increase the level of the neurotransmitter acetylcholine by inhibiting the enzyme acetylcholinesterase from breaking acetylcholine down. Acetylcholine is deficient in Alzheimer's due to the massive death of neurons that contain this neurotransmitter. The cholinergic drugs now on the prescriptive market are: donepezil (Aricept), galantamine (Razadyne), and rivastigmine (Exelon). Common side effects to these cholinergic drugs are nausea, vomiting, diarrhea, indigestion, loss of appetite, and muscle cramps. These bothersome side effects are often not tolerated by a frail patient or one suffering from gastrointestinal problems.

The 4th drug approved for Alzheimer's is memantine (Namenda), a partial blocker of NMDA (N-methyl-D-aspartate) receptors. Overactivation of NMDA receptors is known to occur in a wide range of neurological disorders. NMDA blockers, while useful to reduce excess NMDA receptor activation, can create an overstimulation and excess production of glutamate, an amino acid that can be extremely toxic for brain cells resulting in neuronal damage and death. When blockers of NMDA are not well tolerated by people taking these medications, many side effects can result such as confusion, dizziness, agitation, hallucinations, and coma. Periodic reassessment and careful monitoring by the attending physician are important to avoid as much as possible these side effects when they occur. More recently, a combination drug using memantine and donepezil (Namzaric) was approved by the FDA for moderate to severe Alzheimer's. While the drug combination appeared to be well tolerated when given for several months to Alzheimer's patients, no significant benefits were found when the drug combination was compared to either therapy alone [4].

Medication with these drugs for reducing the symptoms of Alzheimer's has sparked an ongoing controversy between proponents who consider their use as valuable and critics who consider them as harmful and not cost-effective [5]. **The British National Institute for Clinical Excellence has gone on record to criticize the methods and quality of some clinical studies that concluded some efficacy was shown in the use of cholinesterase inhibitors or memantine in Alzheimer's disease and cast doubts that these drugs had the ability to modify quality of life or delay nursing-home placement [6]. These drugs are endlessly advertised on television, newspapers, magazines, and medical journals where they are aimed at consumers and physicians with the clear message that they substantially improve the Alzheimer's patient's life while downplaying the serious side effects they elicit.**

There is a fundamental reason why overtreating a patient with moderate or severe Alzheimer's with these drugs is a questionable exercise that may do more harm than good. The nerve cells that carry and store memories of the past and recent events and which form the framework of our awareness are generally either dead or dying in moderate to severe Alzheimer's. This means that no improvement over the actual mental state is likely to revitalize the cognitive loss. **Thus, it is fair to ask, how will any pharmaceutical drug restore, improve, or mitigate the brain damage that has already been made?**

References

1. Jingami N, Asada-Utsugi M, Uemura K, Noto R, Takahashi M, Ozaki A, Kihara T, Kageyama T, Takahashi R, Shimohama S, Kinoshita A. Idiopathic normal pressure hydrocephalus has a different cerebrospinal fluid biomarker profile from Alzheimer's disease. J Alzheimers Dis. 2015;45(1):109–15.
2. Duthie A, Chew D, Soiza RL. Non-psychiatric comorbidity associated with Alzheimer's disease. QJM. 2011;104(11):913–20.
3. Ryan M, Merrick EL, Hodgkin D, Horgan CM, Garnick DW, Panas L, Ritter G, Blow FC, Saitz R. Drinking patterns of older adults with chronic medical conditions. J Gen Intern Med. 2013;28(10):1326–32.
4. Molino I, Colucci L, Fasanaro AM, Traini E, Amenta F. Efficacy of memantine, donepezil, or their association in moderate-severe Alzheimer's disease: a review of clinical trials. Sci World J. 2013;29(2013):925702.
5. Bond M, Rogers G, Peters J, Anderson R, Hoyle M, Miners A, Moxham T, Davis S, Thokala P, Wailoo A, Jeffreys M, Hyde C. The effectiveness and cost-effectiveness of donepezil, galantamine, rivastigmine and memantine for the treatment of Alzheimer's disease (review of Technology Appraisal No. 111): a systematic review and economic model. Health Technol Assess. 2012;16(21):1–470.
6. Casey DA, Antimisiaris D, O'Brien J. Drugs for Alzheimer's disease: are they effective? Pharm Ther. 2010;35(4):208–11.

Chapter 4
Alzheimer's Then and Now

Alzheimer's Has Been Around Since Ancient Times

Alzheimer's disease has probably existed since mammals roamed the earth although it was not recognized as a disease until 1907. In that year, Alois Alzheimer, a German neuropathologist, published his landmark paper describing a mentally ill patient whose brain after death showed strange deposits later described as senile plaques and neurofibrillary tangles. These microbodies are now known to be the hallmarks of Alzheimer's disease although they are also found in other dementing disorders such as vascular dementia and **dementia pugilistica**, also known as "punch-drunk syndrome," so-called because it was first discovered in boxers who had sustained repeated blows to the head. Repeated concussions to the head is commonly seen not only in boxers but also in military personnel exposed to combat and in athletes playing football, soccer, hockey, and other sports where the head is a target of the game. The preferred term for people who develop cognitive deficits or dementia from these repeated blows to the head is **chronic traumatic encephalopathy (CTE)**.

It is not far-fetched to imagine that if an individual reached the ripe old age of 60 in ancient times and showed signs of Alzheimer's, many cultures would have viewed the strange behavior as a form of demonic possession or religious punishment. This thought is supported by ancient Greek, Roman, and Egyptian writing in the fifth-century BC who described these strange transformations to insanity as a personal madness driven by the gods. The treatments varied depending on the culture and period of time but included incantations, exorcism, bloodletting, starvation, and stoning or beating the afflicted when everything else failed. By the eighth-century AD, things had changed somewhat. In 705, the first psychiatric hospital ward was created in Baghdad, and later, more similar asylums were built in Cairo, Damascus, and Aleppo.

These asylums were basically mental hospices where treatment of inmates consisted of more humane procedures focusing on music, baths, and certain medicines. It

J.C. de la Torre, *Alzheimer's Turning Point*,
DOI 10.1007/978-3-319-34057-9_4

is hard to figure out where or how Alzheimer's patients may have been treated from the eighth to the nineteenth century because many of the symptoms relating to depression, schizophrenia, memory loss, and manic behavior were lumped together as one entity described as so many facets of lunacy. Diabolical and divine causes for madness in Europe during the middle ages were thought to result from a malfunction of 4 humors: black bile, yellow bile, phlegm, and blood. Drilling a hole in the skull to let out these humors out was often the first choice of treatment.

This absence of differentiating one mental disorder from another gave rise to lunatic asylums that proliferated in the 18th and 19th century where the primary treatment was brutal restraint and containment using leg irons, chains, and manacles. These asylums grew at a blurring fast pace due to the rising populations in Europe and Asia. Nevertheless, in some centers, harsh treatment of the mentally ill began to change somewhat around 1796 when **William Tuke**, a Quaker businessman, established The York Retreat, located in York, England.

York Retreat operated as a new concept in the treatment of the insane. The retreat did away with the leg irons and manacles and replaced these with kindness, attention, religion, and order with the ever-present conviction that inmates could one day recover their sanity. This philosophy led to an improved self-image among the inmates and better management of their needs since it was determined that fear and abuse worsened insanity. The York Retreat soon became a model for other asylums in the world and especially in late eighteenth-century Paris where **Philppe Pinel** adopted some of Tuke's methods banning physical abuse and manacles from La Bicêtre mental hospital and later at the Salpêtrière hospital which he directed. Pinel's only departure from Tuke's approach of the insane was to dismiss religious practice from his treatments which Pinel felt made insanity worse by instilling self-reproach and guilt feelings.

One of William Tuke's relatives in the nineteenth century was one of Europe's most well-known psychiatrists, a man whose name was oddly enough quite appropriate for his profession, **John Batty Tuke**.

Batty Tuke, as he is historically known, plays a significant role in our Alzheimer's story because he not only treated the insane humanely at several Edinburgh asylums where he worked but also wrote extensively about mental illness in the late nineteenth century, focusing on the breakthrough idea that **physical disease is the cause of mental illness**. This revolutionary concept departed from the long-held belief that insanity was a demonic or heavenly punishment or even a psychological perversion of the intellect. Instead, Batty Tuke stressed the importance of brain anatomy, pathology, and physiology as a way to understand mental disorders. The concept that a vascular component could be associated with the development of dementia probably began in 1873, when Batty Tuke did report **unusually distorted large blood vessels in the brains of mentally disturbed patients who had died** [1]. Neither Batty Tuke nor anyone else may ever know whether some of the brains he examined with brain vessel distortions belonged to Alzheimer's patients. Such a detail could have influenced Alois Alzheimer to look beyond the plaques and tangles that he astutely described in the brain slices of his mentally ill patient Auguste D after her death [2].

Twisted Blood Vessels

Vessel distortions in the brains of the mentally ill were later confirmed in microscopic studies by several investigators where it was shown that microvessels underwent kinking, twisting, and tortuous patterns [3–5]). These were mainly anatomic observations that did not include a physiological explanation. In 1990, **Dr. V. Fischer** and colleagues at the St. Louis School of Medicine noted similar microvessel distortions in the hippocampal brain region of Alzheimer's patients who had died [6]. This was an important observation because the hippocampus is a center known to have a pivotal role in memory and learning and one of the first targets of neurodegeneration seen in Alzheimer's brains. Fischer observed not only vessel distortions in the Alzheimer's brains but also a reduction in the number of these microvessels in key brain areas known to be associated with cognitive function [6]. These investigators assumed that the microvessel distortions and reduced number of vessels resulted from a "pathologic factor" intrinsic of Alzheimer's dementia since the brains of cognitively intact individuals did not exhibit such pathology. Fischer speculated that Abeta protein lining the inside vessel walls of Alzheimer's brains may have crossed into the brain tissue causing toxic damage to brain cells and microvessels.

In 1989, **Dr. Arnold Scheibel** and his group [7] at UCLA, extended Fischer's observations using scanning electron microscopy on Alzheimer's brain slices which greatly magnified the vessel structures. Scheibel noted seeing irregular pouches and distinct outgrowths on the cortical vessels of one Alzheimer patient at autopsy. He described these outpouchings as showing a "lumpy-bumpy" appearance [7]. Scheibel speculated as Fischer had done before him that the vessel distortions could damage the blood–brain barrier in Alzheimer's brain leading to the penetration of abnormal matter into the brain tissue that could then contribute to the characteristic Alzheimer's pathology [7]. The blood–brain barrier is a type of security system or toll gate located in brain capillaries that allows the entry of essential nutrients while keeping out everything else. Its disruption can cause toxins to enter the brain and kill neurons. Although this could potentially damage brain tissue, there is no convincing evidence that such a process occurs in Alzheimer's as the cause of neurodegeneration. There is evidence that amyloid beta enters the brain through a faulty blood–brain barrier to produce cerebral amyloid angiopathy which would be an additional burden to lower cerebral blood flow due to partial blockade of the vessel lumens. The mechanism for this process is not entirely clear.

Cerebral Hypoperfusion Causes Alzheimer's

A first principle in biology states that in animals with a circulatory system, when a blood vessel stops flowing, the cells it nourishes quickly die. This principle applies to all living cells served by blood vessels.

In 1993, after several years of studying chronic brain hypoperfusion in young and old rats, we came to the conclusion that reduced cerebral blood flow in older rats was the cause of cerebral microvessel damage, memory loss, and neuron death that began primarily in the hippocampus then extended to cortical regions [8] (see also Chap. 14). This would later be known as the **vascular hypothesis of Alzheimer's disease**. At that time we had already established in previous studies that normal older rats had lower brain blood flows than normal young rats. Thus, when cerebral blood flow was experimentally reduced in young and old rats, only the older animals revealed consistent, memory impairment, indifferent exploratory cage behavior (apathy), and absence of grooming behavior after several weeks. Since no small or major strokes had been detected postmortem in any of the animals' brains, the memory deficiency indicated that poor blood flow was the principal source of the cognitive impairment. Only after 6–12 months of cerebral hypoperfusion in the old rats was neuronal loss, protein abnormalities, ventricular enlargement, and mediotemporal brain atrophy noted [9]. These structural and metabolic changes in the old, hypoperfused rats were characteristically seen in human Alzheimer's disease. The clinical picture of brain damage in the hypoperfused old rats resembled that of severe Alzheimer's brain. (Figs. 1.4 and 4.1)

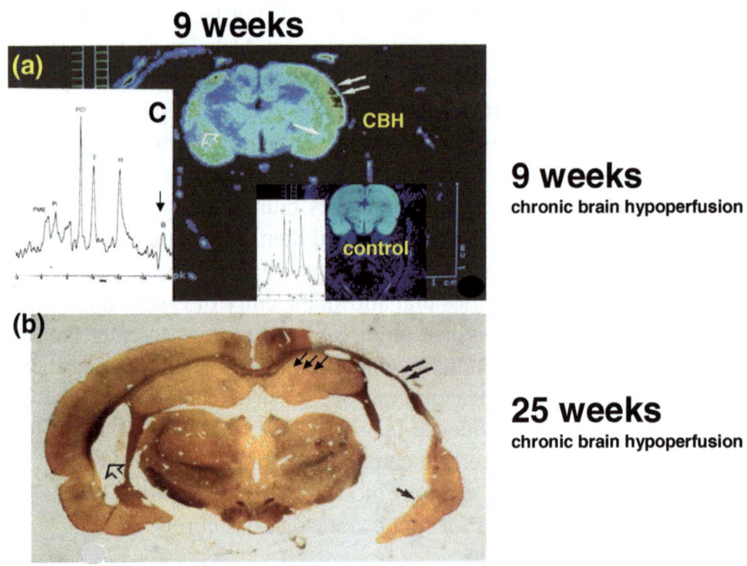

Fig. 4.1 Old rat brain subjected to chronic brain hypoperfusion (*CBH*) for 9 weeks examined with magnetic resonance imaging (**a**). *Arrows point* to *yellow-dark* tissue where blood flow is poor. However, no neurodegenerative changes are seen at this time much like the non-hypoperfused control. ATP energy reduction after 9-week CBH is indicated by depressed peak on magnetic resonance spectroscopy (*arrow* **c**) but not control. Same brain on week 25 **b** shows severe atrophy of parietal cortex (*double arrows*), temporal cortex (*arrow*), and hippocampus (*triple arrows*) mimicking Alzheimer's pathology. *Open arrow* shows ventricle enlargement. Severe memory loss characterized these older animals. Compare this rat image with of the severe atrophy seen in Alzheimer's brain (Fig. 1.4). Adapted from de la Torre et al [9].

In the old, hypoperfused rats, however, no plaques or tangles were seen unlike the human Alzheimer's brain. The realization that marginal brain blood flow typically seen in normal older rats and normal elderly humans could be made worse by a mechanical or physiological process resulting in memory impairment and eventual nerve cell death was one of the happiest observations this author ever made. **It meant that possibly Alzheimer's disease was a disorder that affected older people whose brain blood flow was already marginal from aging but somehow worsened through the appearance of "unknown factors" to induce severe cerebral tissue damage and cognitive deterioration**. This concept broke with the classical thinking that neurodegeneration uniquely stoked the production of vascular pathology seen in dementia brains. **It suggested instead, as we proposed in 1993, that Alzheimer's disease was a vascular disorder with neurodegenerative consequences**, not the other way around [8]. That proposal meant that if triggers associated with the vascular perfusion abnormalities could be found, Alzheimer's disease might be prevented or its deterioration substantially slowed down by targeting those 'unknown' triggers. The burning question still remained in 1993, where or what were the triggers that pulled down cerebral perfusion to abnormal levels in Alzheimer's brain?

Challengers of the vascular hypothesis of Alzheimer's disease pointed out that cerebral hypoperfusion associated with this dementia was due to dead or dying neurons which no longer required the metabolic energy supply needed to stay alive. However, studies have shown that this conclusion is faulty. Neurochemical and neuroimaging findings indicate that **cortical hypoperfusion in Alzheimer's exceeds the reduction in metabolic demand for oxygen** and is therefore a primary pathological initiator of reduced blood flow [10, 11]. The reason being that if brain hypoperfusion were the result of less oxygen requirement needed by dying or dead neurons, the regional oxygen demand would be unchanged which in the case of Alzheimer's, it is not. Brain hypoperfusion results in "misery perfusion" where the regional metabolic demand for oxygen is reduced due to narrowing or occlusion of vessels supplying the brain with blood flow not because of dying or dead neurons. Misery perfusion in Alzheimer's patients has been detected in the form of **increased regional oxygen extraction fraction** using positron emission tomography (PET) scans [12]. When neurons are faced with brain hypoperfusion, they have the ability to extract more oxygen from blood microvessels in order to prevent tissue hypoxia, functional impairment, or hemodynamic compromise. This is reflected when the oxygen extraction fraction is increased. However, there may be a limit to how much oxygen neural tissue may be able to extract from the blood supply, a situation that may trigger neuronal dysfunction leading to cognitive decline.

Other studies have reported that cerebral hypoperfusion is present in cognitively normal, aged individuals who later develop Alzheimer's [13, 14]. These findings imply that if neurodegeneration were the driving force behind cerebral hypoperfusion, one would expect cognitive deterioration due to neurodegenerate changes to be present prior to such hypoperfusion, but evidence indicates brain hypoperfusion precedes cognitive decline and neurodegeneration [12, 13].

In 1994, we reported our suspicion that certain vascular-related conditions which were known then to reduce blood flow to the brain seemed like prime contributors to initiate the lethal evolution of Alzheimer's dementia [15]. The list of known vascular risk factors for Alzheimer's was limited in 1994 but later grew rapidly so that by 1997, a dozen additional risk factors, including hypertension, atherosclerosis, and smoking, were reported [16]. Thus, the triggering factors to cerebral hypoperfusion were now being quickly discovered and epidemiologically linked to Alzheimer's. At the present, several dozen vascular risk factors for Alzheimer's are known [17]. This intriguing part of the story is more fully discussed in Chap. 12.

The presence of cerebral hypoperfusion has also been identified in brain scans prior to mild cognitive impairment, a preclinical stage that often converts to Alzheimer's [13]. Neuroimaging of patients with mild cognitive impairment shows a consistent reduction of regional blood flow several years before Alzheimer's symptoms appear. [18, 19] More recently, it now appears that Alzheimer's conversion from mild cognitive impairment can be predicted in cognitively normal individuals by evaluating white matter lesions in the brain of elderly persons [20]. Subjective cognitive decline is a common manifestation of old age and a concern of elderly people who perceive the reduced cognitive function as a warning of impending dementia despite the fact that no **objective** cognitive decline is present. Nonetheless, it has been shown that in the presence of brain lesions in the white matter (called white matter hyperintensities), subjective cognitive decline may sometimes represent a prodromal stage of mild cognitive impairment, a common precursor of Alzheimer's [20].

Substantial evidence now indicates that chronic brain hypoperfusion is the main precursor of Alzheimer's and that conditions which create poor blood flow to the older brain appear to set the stage for the cognitive decline that characterizes preclinical Alzheimer's [8–24] Some of these conditions may take decades to alter cognitive function. They include the following:

1. Distorted brain microvessels;
2. Atherosclerosis of large brain arteries;
3. Atherosclerosis of peripheral arteries supplying the brain;
4. Low cardiac output;
5. Aortic and arterial stiffening of vessels supplying blood flow to the brain;
6. Brain microvessel thrombi or hemorrhage;
7. Vascular risk factors to Alzheimer's (see discussion, Chap. 12).

The intriguing question can now be posed: Are depression, mood swings, anxiety, and other dysemotional states that often accompany Alzheimer's also due to poor blood flow to the brain? The answer may open a spectrum of therapeutic opportunities that will change how medicine views and manages these ailments.

References

1. Tuke JB. On the morbid histology of the brain and spinal cord as observed in the insane. Br For Med Chir Rev. 1873;51:450–460.
2. Alzheimer A. Uber eine eigenartige Erkrankung der Hirnrinde. Allgemeine Zeitschrift für Psychiatrie und phychish-gerichtliche Medizin, (Berlin). 1907;64:146–148.
3. Ravens JR. Vascular changes in the human senile brain. In: Cervos-Navarro J, editor. Pathology of Cerebrospinal Microcirculation. New York: Raven Press, 1978: pp. 487–501.
4. Hassler O. Arterial deformities in senile brains. Acta Neuropathol. 1967;8:219–29.
5. Bell MA, Ball MJ. Morphometric comparison of hippocampal microvasculature in ageing and demented people: diameters and densities. Acta Neuropathol. 1981;53(4):299–318.
6. Fischer VW, Siddigi A, Yusufaly Y. Altered angioarchitecture in selected areas of brains with Alzheimer's disease. Acta Neuropathol. 1990;79:672–9.
7. Scheibel AB, Duong TH, Jacobs R. Alzheimer's disease as a capillary dementia. Ann Med. 1989;21(2)
8. de la Torre JC, Mussivand T. Can disturbed brain microcirculation cause Alzheimer' disease? Neurol Res. 1993;15:146–53.
9. de la Torre JC, Fortin T, Park GA, Butler KS, Kozlowski P, Pappas BA, de Socarraz H, Saunders JK, Richard MT. Chronic cerebrovascular insufficiency induces dementia-like deficits in aged rats. Brain Res. 1992;582(2):186–95.
10. Thomas T, Miners S, Love S. Post-mortem assessment of **hypoperfusion** of cerebral cortex in Alzheimer's disease and vascular dementia. Brain. 2015;138(Pt 4):1059–69.
11. Miners JS, Palmer JC, Love S. Pathophysiology of **Hypoperfusion** of the Precuneus in Early **Alzheimer's Disease**. Brain Pathol. 2015;Oct 9.
12. Nagata K, Kondoh Y, Atchison R, Sato M, Satoh Y, Watahiki Y, Hirata Y, Yokoyama E. Vascular and metabolic reserve in Alzheimer's disease. Neurobiol Aging. 2000 Mar–Apr;21 (2):301–7.
13. Xekardaki A, Rodriguez C, Montandon ML, Toma S, Tombeur E, Herrmann FR, Zekry D, Lovblad KO, Barkhof 615 F, Giannakopoulos P, Haller S. Arterial spin labeling may contribute to the prediction of cognitive deterioration in healthy elderly individuals. Radiology 2–14;274, 490–9.
14. Ruitenberg A, den Heiker T, Bakker SL. Cerebral hypoperfusion and clinical onset of dementia: the Rotterdam study. Ann Neurol. 2005;57:789–794
15. de la Torre JC. Impaired brain microcirculation may trigger Alzheimer's disease. Neurosci Biobehav Rev. 1994;18(3):397–401.
16. de la Torre JC. Hemodynamic consequences of deformed microvessels in the brain in Alzheimer's disease. Ann N Y Acad Sci. 1997;826:75–91.
17. de la Torre JC. Is Alzheimer's disease a neurodegenerative or a vascular disorder? Data, dogma, and dialectics. Lancet Neurol. 2004;3(3):184–90.
18. Huang C, Wahlund LO, Svensson L, Winblad B, Julin P. Cingulate cortex hypoperfusion predicts Alzheimer's disease in mild cognitive impairment. BMC Neurol. 2002;2:9.
19. Lee SJ, Ritchie CS, Yaffe K, Stijacic CI, Barnes DE. A clinical index to predict progression from mild cognitive impairment to dementia due to Alzheimer's disease. PLoS ONE. 2014;9 (12):e113535.
20. Benedictus MR, van Harten AC, Leeuwis AE, Koene T, Scheltens P, Barkhof F, Prins ND, van der Flier WM. White matter hyperintensities relate to clinical progression in subjective cognitive decline. Stroke. 2015;46(9):2661–4.
21. Laosiripisan J, Tarumi T, Gonzales MM, Haley AP, Tanaka H. Association between cardiovagal baroreflex sensitivity and baseline **cerebral** perfusion of the hippocampus. Clin Auton Res. 2015;25(4):213–8.

22. Roher AE, Esh C, Kokjohn TA, Kalback W, Luehrs DC, Seward JD. Circle of willis atherosclerosis is a risk factor for sporadic Alzheimer's disease. Arterioscler Thromb Vasc Biol. 2003;23:2055–62.
23. Dai W, Lopez OL, Carmichael OT, Becker J, Kuller LH, Gach HM. Mild cognitive impairment and alzheimer disease: patterns of altered cerebral blood flow at MR imaging. Radiology. 2009;250:856–66.
24. Alsop DC, Detre JA, Grossman M. Assessment of cerebral blood flow in Alzheimer's disease by spin-labeled magnetic resonance imaging. Ann Neurol. 2000;2000(47):93–100.

Chapter 5
Unproven Hypotheses on the Cause of Alzheimer's

The Abeta Hypothesis: A Die-Hard Concept

The best way to tell the story of the Abeta hypothesis is to start with a concept that logically should have clearly explained the cause of Alzheimer's disease 20 years ago, but did not.

Before plunging into the many crevices that envelop the Abeta hypothesis, it is noteworthy to point out that when a widely held medical belief is shown to be wrong, proper treatment and management of patients associated with that belief will be less than desirable or conspicuously absent. Scientific proof in biology is generally difficult to demonstrate. **One factor that strengthens a hypothesis is the amount of verifiable evidence available to support it**. Basic evidence gathered from test tubes or animal experiments is usually the starting point for clinical studies in humans, assuming the therapy or intervention is proven safe. However, even when an intervention is highly effective to experimental animals for alleviating or reversing an induced lesion or disease, if it does not work in human clinical trials, as is often the case, the experimental intervention is abandoned and the science moves on to examine other ideas.

Consequently, even powerful experimental evidence (i.e., strongly suggestive) is not proof that something is true. When clinical evidence is lacking or absent with respect to disease's causative factors, it is critically important for scientists and for science not to latch on to this "failed hypothesis" for the sake of convenience or financial reward but to actually seek other explanations and move the research elsewhere. But, what happens when this logic is not followed?

According to the science historian **Thomas Kuhn**, it is axiomatic that most scientists with an intellectual or financial stake in a theory tend to ignore the facts that may undercut their views [1]. Kuhn also astutely observed,

J.C. de la Torre, *Alzheimer's Turning Point*,
DOI 10.1007/978-3-319-34057-9_5

> No part of the aim of normal science is to call forth a new sorts of phenomena; indeed those that do not fit the box (containing the prevailing paradigm) are often not seen at all. Nor do scientists normally aim to invent new theories and are often intolerant of those invented by others [1]

Kuhn correctly predicted how the majority of researchers embraced a hypothesis 20 years ago concerning the cause of Alzheimer's at the expense of any other idea or concept. These scientists became fully invested in studying a hypothesis based on what they saw as the "toxic" brain formation of plaques and tangles routinely seen in many (but by far, not all) patients that died with this dementia. Any other theory formulated within the last two decades that did not compliment the plaque and tangle hypothesis became an annoying distraction to its supporters. Moreover, the plaque and tangle hypothesis, which later mutated into the so-called **Abeta hypothesis**, was strongly promoted and generously funded by several pharmaceutical companies who envisioned big profits from a proliferation of Abeta projects leading to anti-Abeta products. Many careers and professional reputations were built around the Abeta hypothesis, and this loyalty to an unproven proposal became the prevailing paradigm and gold standard for dementia research that continues to this day.

To be fair, pharmaceutical and government funding of the Abeta hypothesis 20 years ago was critical in keeping many researchers from shutting down their laboratories for lack of funding. This money was needed to pay their own salary, their staff, buy needed equipment and supplies, and pay the rent at their institutions. The Abeta hypothesis thus became the accepted paradigm for nearly all Alzheimer's research in the world and flourished as a dominating force at medical conferences and publications in scientific journals. In the early 1990s, the Abeta hypothesis also spotlighted the commonness of Alzheimer's as a pathological disorder of the elderly and pointed out the devastating consequences associated with this dementia. The intense research of cell and molecular abnormalities found in Alzheimer's brains and experimental models stimulated a mother lode of information of a cognitive disorder that had been ignored for nearly a century.

Thus, money from the pharmaceuticals supported many non-profit Alzheimer's agencies, as well as experienced academic and new investigators, their graduate students, and their staff. The only catch to such money was that the research performed needed to compliment the drugmakers' goals and objectives.

A Seductive Hypothesis

On the negative side, the Abeta hypothesis was beguiling in catching the collective imagination of many investigators who, with the passage of time and pharmaceutical strong-arm tactics, came to dominate peer review of scientific articles, funding from granting bodies and most of Alzheimer's conference proceedings. Pharmaceutical money kept an almost impregnable citadel jealously guarded around the Abeta hypothesis from most scientific criticisms or detractors. This was a testament to the pharmaceutical power and willingness to protect their multi-billion-dollar investment despite growing clinical evidence that suggested the Abeta hypothesis was dismally weak.

The Abeta hypothesis is primarily based on three linked assumptions, briefly stated: (1) **Familial Alzheimer's** disease is driven by genetic missense mutations of the amyloid precursor protein (APP) on chromosome 21 and of presenilins (PS1 and PS2) on chromosomes 1, 14. APP mutation is said to generate excessive production of $Abeta_{42}$, a supposedly "toxic" peptide that causes familial Alzheimer's disease; (2) by extrapolation, the abnormal process responsible for the overproduction of Abeta in familial Alzheimer's disease mimics the pathophysiology of the non-genetic form and consequently causes **sporadic Alzheimer's**; and (3) clearance or abolishment of such Abeta-containing plaques in Alzheimer's brain will improve cognition and favorably alter the neurodegenerative progression of this disease [2].

This syllogistic argument leans heavily on its major premise that familial Alzheimer's disease (which affects less than 3 % of all Alzheimer cases) is caused by toxic Abeta accumulation in the brain, which even if one accepts as valid, and this is debatable [3, 4], it does not follow that the genetic and non-genetic forms of Alzheimer's express the same pathology since they do not share the aberrant genes supposedly responsible for familial Alzheimer's disease.

Without such evidence, the major premise that the genetic cause of Alzheimer's is also the cause of sporadic Alzheimer's would not appear to provide the needed degree of support for that conclusion and instead introduces a serious error in deductive reasoning.

Logically, if the Abeta argument is to make any sense, the dementia that develops in sporadic Alzheimer's should also show the same gene mutation as familial Alzheimer's disease because without such mutation there would be no familial Alzheimer's disease! Consequently, there is no evidence that gene mutations are involved in sporadic Alzheimer's. If only this much information were available to describe the misgivings of the Abeta hypothesis, it would have been enough for most investigators to walk away from this proposal in search of other research opportunities.

However, this curious ambiguity was put on hold as several pharmaceutical companies expressed a financial interest in mining the Abeta hypothesis, prompting many investigators to hang on and study the problem deeper. Since the Abeta promoters knew that no gene mutation of APP occurred on chromosome 21 in sporadic Alzheimer's, they concluded, with pharmaceutical help, that the excess brain formation of Abeta must be due to other sources, the gist of which form the basis for over 20,000 scientific papers in the last two decades. Collectively, these papers have yet to pin down the exact pathological process that involves APP in sporadic Alzheimer's.

Other difficulties with the amyloid hypothesis began to appear in the scientific literature. These difficulties can be summarized as follows: (1) Abeta deposition in the brain does not relate to dementia severity [3]. (2) Many patients without dementia have the same density of senile plaques as patients with AD [4]. (3) Amyloid deposition is not the earliest neuropathological event observed in those afflicted with the disease [5]. (4) Many cognitively healthy elderly people have abundant senile plaques in their brains but no signs of AD [6]. (5) Amyloid deposition in the brain does not correlate with neuronal, metabolic, or synaptic loss [7]. (6) Amyloid plaques can be found in other dementias including vascular dementia. (7) Experiments with transgenic mice that produce Abeta deposits in the brain show, as in human beings, no relation between such deposits and neuronal, metabolic, or synaptic loss [8]. Finally, the Abeta hypothesis stipulates that Abeta accumulation in brain causes neurofibrillary changes, a pathological process found in dead or dying neurons. But evidence indicates that Abeta deposition **rarely occurs in the early stages of Alzheimer's in the basal nucleus of Meynert where many neurons are seen dead or dying.** The basal nucleus of Meynert is a targeted brain region for very early and consistent neurofibrillary changes in Alzheimer's, a finding indicative that Abeta is not the cause of neurofibrillary pathology and therefore not the cause of Alzheimer's neuropathology [9].

Another incongruity concerning Abeta deposition as the cause of Alzheimer's is the fact that Abeta brain accumulation is also seen in other neurodegenerative disorders such as Parkinson's disease, vascular dementia, and Creutzfeldt–Jakob disease. Creutzfeldt–Jakob disease by the way is known to be caused by a misfolded protein called prion [10]. This fact alone indicates that Abeta is a pervasive pathological product of neurodegeneration rather than the cause of dementia (Fig. 5.1).

To skate around these difficulties, proponents of the original amyloid hypothesis have recently suggested that **soluble oligomers of Abeta$_{42}$** and not its insoluble amyloid fibrils or monomers, as previously suggested, are the toxic cause of synaptic dysfunction in Alzheimer's and also in animal models that mimic Alzheimer's. Soluble oligomers are misfolded proteins considered to be neurotoxic to neurons and supposedly responsible for the progressive cognitive decline in

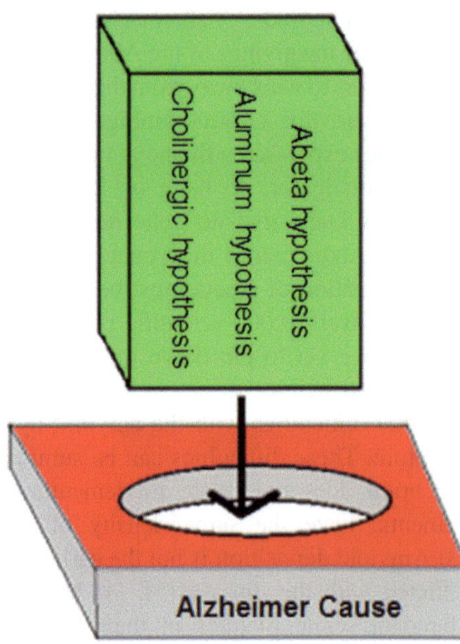

Fig. 5.1 The square peg into a round hole dilemma illustrates the tenacity, persistence, and futility of holding on to a hypothesis that has consistently failed to deliver the evidence required to support its prediction. See text for details

Alzheimer's disease. The only solid evidence for this conclusion is based not on human brain but on genetically manipulated "transgenic" mice bred to accumulate these abnormal proteins in their brain [11].

Soluble oligomers vary in size and morphology and appear to represent temporary intermediates in the pathway of fibril formation. Some forms of this molecule are highly toxic when exposed to cell cultures in vitro. When they are manufactured outside the brain cells, as they are presumed to do in Alzheimer's brains, it is speculated that they could introduce toxic changes in the cell membrane by allowing excessive calcium ions to enter the cell and kill or damage the cell, a theory strongly promoted to be the cause of Alzheimer's [11].

When a Hypothesis Does Not Work

The "soluble oligomer hypothesis" on the cause of Alzheimer's has led to a legion of studies in rodents genetically engineered to mimic Alzheimer's pathology. The reason for this renewed interest in the Abeta hypothesis is the link between soluble oligomers and insoluble fibrils that can cause plaque formation in Alzheimer's dementia and also because soluble oligomers are believed to show better correlation with the degree of cognitive loss in patients than the presence of fibrillar insoluble forms.

There is scant characterization of Abeta-soluble oligomers concerning where and how they develop in human brain tissue, where they accumulate to exert their alleged cytotoxicity, and whether they are the product of some specific metabolic aberration. Abeta-soluble oligomer toxicity is derived mainly from in vitro evidence. This fact poses a red flag of caution in extrapolating the results and conclusions of in vitro studies to in vivo reality.

No extraction of oligomers to characterize their structure in Alzheimer's has been done to this day, and most of the information regarding their toxicity in brain tissue is derived from artificial preparations in mice made to harbor familial Alzheimer's disease–APP mutations. Abeta oligomeric toxicity in mice genetically engineered shows only a killing attack on cultured neurons but no frank neuronal loss when these soluble oligomers are exposed to mice brain despite the formation of many Abeta plaques [12].

Even the toxicity of Abeta found in Alzheimer's brains has been questioned. **Dr. Nikolaos Robakis,** a researcher at New York's Mt. Sinai School of Medicine, reported that no toxic soluble oligomers specific to Alzheimer's have been found and that amyloid plaque depositions could not explain the neurodegeneration typical of dementia [13]. Robakis showed that (1) amyloid depositions at levels similar to those seen in Alzheimer's brain are often detected in normal individuals; (2) mouse models genetically manipulated to express high amounts of brain amyloid failed to show significant cerebral neurodegeneration [13]. Moreover, the genetic manipulation of these mouse models is dramatic, because it contains 10,000 times higher the amount of Abeta in their brains as compared to human Alzheimer's brain [13].

Despite all this evidence, there has been a rush of basic studies focusing on the assumed actions, functions, and role of Abeta oligomers on Alzheimer's during the last decade, but most of the knowledge gained so far remains superficial, at best. As any curious scientist might conclude, it would be very nice to find out in the next decade whether these oligomers have any role in the development of dementia, but, in the meantime, should we not try to impact Alzheimer's in a more decisive manner and in considerably less time?

Are we ready to therapeutically target the elusive Abeta oligomers as was done by pharmaceuticals for the fibrillar forms of this molecule? If so, history may teach a lesson on how hundreds of millions of dollars were misspent on more than a dozen human clinical trials to prove that active immunization against Abeta plaque formation in Alzheimer's patients was the cure we were all hoping for. Nonetheless, that is not what happened.

Consistent Clinical Trial Failures

Often, positive animal experimentation is sufficient to warrant human clinical trials after which it can be determined whether the experimental intervention is safe and effective. The dictum that scientific theories can be made or broken by experimentation is well known in research. **Experimentation using the scientific method *requires the elimination of a hypothesis* if experiments or clinical trials repeatedly contradict its hypothetical predictions**. This axiom was defied by the proponents of the Abeta hypothesis repeatedly. For example, the Abeta hypothesis predicted that reducing or clearing amyloid beta-containing plaques from Alzheimer's brains should improve cognitive function and slow down or arrest the progressive pathological process. Repeated clinical trials showed neither prediction occurred [14]. These clinical trials included active and passive Abeta immunotherapies of such drugs as Neurochem's **AlzheMed [tramiprosate**], Myriad Genetics' **flurizan [tarenflurbil**], Elan's **AN-1792 and 301**, Pfizer's **bapineuzimab,** and Lilly's **solanezumab** [15–18]. The drugs' common goal was to wipe out the amyloid plaque deposits found in Alzheimer's brain.

In one trial using Elan's AN1792, immunization with Abeta42 resulted in virtually complete amyloid plaque removal from the brain of Alzheimer's patients, but despite this, progressive neurodegeneration was not prevented or alleviated [15]. In a subsequent trial, bapineuzimab was retested in Alzheimer's patients, and after 78 weeks, no significant differences in the rate of cognitive decline were found when compared to placebo-treated controls. Nearly 10 % of bapineuzimab-treated patients developed vasogenic edema (brain swelling) during the trial. Thus, not only these trials failed to improve cognition or alter neurodegenerative progression of Alzheimer's, but also some of these studies had to be stopped because patients taking the anti-Abeta treatments developed severe adverse events [19].

One major concern about the Abeta hypothesis is the design of clinical trials to address answers rather than assumptions, and as such, evidence was gathered to prove these answers or justify their failure. This is antithetical to the scientific method. Nevertheless, some investigators dismiss the idea that repeated clinical failures that have tested the Abeta hypothesis should not be taken seriously until an extra 5-year retesting of anti-Abeta therapy is performed and in the event that clinical failures continue, the testers should then “reorganize their mind-set” [20].

This mind-set has not yet affected the pharmaceutical technocrats and investigators involved in the clinical design of these trials. The tactic after each negative trial included a scramble to offer possible reasons for the failure followed by the resetting of new trials using either identical or me-too copies of the anti-Abeta treatments. The pharmaceuticals involved in this deception are, to paraphrase H.L. Mencken, **swathing the ugly facts in bandages of soft illusion**. In my judgment, Mencken’s comment may be too charitable in describing big pharma’s real motives.

Smothering the Think Tank

Thus, the continued reinvention of these anti-Abeta compounds continues to be retested on other Alzheimer’s patients in several ongoing trials. Why do these pharmaceuticals persist in clinically retesting the same failed concept over and over again and expecting a different result? In the case of the Abeta hypothesis, the answer is money. Pharmaceutical industry executives have learned the calculus of profit and made that their corporate mantra, while in the process, any groundings of human decency or concern for human suffering have been sidestepped or ignored.

Curiously, the pharmaceuticals involved in these anti-Abeta treatments claim they spend hundreds of millions of dollars as an investment to target a multi-billion-dollar industry. The multi-billion-dollar target in this case is convincing the FDA that their product is effective and safe. Neither goal has been achieved, possibly because of the following findings:

(1) A number of studies have indicated that substantial Abeta burden can be found in cognitively intact older people even though the plaque and tangle distribution, density, and topographical progression of these “hallmark” deposits are the same as that seen in symptomatic Alzheimer’s patients [3, 6, 21]. (2) Two neuroimaging studies have confirmed that the amount of plaque deposition in cognitively normal and Alzheimer’s brain is similar [22, 23]. In a large community-based necropsy study of elderly patients’ brains aged 70–103, 33 % of normal, non-demented individuals showed similar Abeta plaque densities and deposition as those with Alzheimer’s [22].

More recently, a second study showed that 33–65 % of cognitively normal persons who underwent (11) C-Pittsburgh compound B (PiB) scans (a technique that detects Abeta brain deposits) had similar high PiB binding as patients diagnosed with Alzheimer’s [23]. (3) Abeta plaque formation in Alzheimer’s brain is a downstream pathological event, meaning it occurs after the disease has evolved [5], a finding which explains why significant Abeta brain accumulation does not appear to be associated with worsening of cognitive function. Thus, v**irtual clearing of**

amyloid plaques from human brain with anti-Abeta immunization does not prevent progressive neurodegeneration nor the advancing cognitive loss typical of Alzheimer's [15].

Consequently, amyloid deposition does not correlate with the neurodegenerative process that includes neuronal, synaptic, and metabolic loss, nor does it correlate with the severity of dementia. The conclusion from these findings is that Abeta elimination from the brain does not improve any of the features that characterize Alzheimer's dementia. The clinical evidence thus far has revealed that the cause of Alzheimer's by Abeta overproduction in the brain has not been demonstrated. Can the field of Alzheimer's research keep pretending that it makes sense to continue forcing a square peg into a round hole? (Fig. 5.1)

At this point, the reader may ponder as to why create such a big deal about whether Abeta has a genetic basis or not. If the counterargument to the Abeta hypothesis were about selling bubble gum or men's perfume, unconcern might be understandable. Unfortunately, the topic of Alzheimer's and its possible cause are directly related to the therapeutic prevention of its risk factors and treatments not yet developed. Knowing the correct cause of Alzheimer's is primarily about people's health and well-being. As such, the dreadful misery of Alzheimer's disease causes its victims and the hardships that fall on its caregivers should energize scientists and clinicians to seek a research approach that will significantly reduce or ideally eliminate this formidable disorder. Clinical trials and old concepts that have continually failed in the past should be abandoned to make room for new ideas that can offer hope to those not yet struck by this tragic disorder.

The collective rallying around an unproven hypothesis for two decades has neatly smothered any other proposal that might provide a clearer understanding of how to reduce the prevalence of this dementia. The Abeta hypothesis also propelled a number of investigators to become very rich when they created start-up companies that later were sold to major pharmaceuticals for towering profits. This get-rich quick entrepreneurship by some leading scientists would have been tolerable if it had been based on a significant invention or discovery that directly helped patients. However, in this case, it amounted to outright fraud [24].

Scientists receiving substantial consulting money from a pharmaceutical to expound the virtues of a product favorable to that pharmaceutical often do not mention in their articles this conflict of interest, despite ethical rules from reputable journals that demand it. When authors submit a manuscript to a scientific journal for publication, they are responsible, according to the International Committee of Medical Journal Editors, for disclosing all financial and personal relationships, including consultation fees that can bias or be perceived to bias their work. However, a cozy tit-for-tat arrangement between a scientist and big pharma to make serious money from this rarely enforced rule is frequently ignored when scientific articles are submitted to a journal. Well-known Alzheimer's researchers of this ethical breach have been identified but rarely punished when found out [25]. When financial gain takes precedence over patient welfare, the offender not only offends the profession but also shows no regard for the patients' suffering or their caregivers. This is baffling and loathsome to many other decent scientists when this type of corruption occurs.

Consequently, start-up companies set up for taking advantage of dubious scientific assumptions which are later sold at sky-high prices to colluding pharmaceuticals do nothing to provide hope or understanding to those facing the hard reality of Alzheimer's disease.

Not to be outdone, other investigators who joined the Abeta club have also been boosting their salaries by doing consulting work, joining pharmaceutical advisory boards or directing Abeta clinical trials for drug companies. Little sense of morality ever enters these *quid pro quo* arrangements with big pharma [24]. To say that money is the principal motive in aggressively pushing the Abeta hypothesis on all fronts would not be an overstatement.

The most disturbing aspect of the Abeta hypothesis is that its main advocates rarely if ever seem to consider other hypotheses that might better explain the cause of Alzheimer's either in their conference presentations or in their scientific publications. Vascular risk factors, cell cycle abnormalities, or oxidative stress are ignored as potential constructs that may better explain the pathological evolution and clinical management of dementia. **The Abeta overlords and their followers have created an enormous fishbowl without the fish where everyone can admire from afar the colorful flashes of light caught in a beam of sunlight glancing off the imaginary aquarium.**

Ironically, even though the Abeta hypothesis appears to be in a death spiral as far as a revered paradigm, it still keeps a footprint in all major Alzheimer's conferences, publications, and competitive funding. For decades, the Abeta hypothesis has been sucking out the oxygen from the Alzheimer's think tank with respect to funding of novel research projects while keeping a stranglehold on publications and scientific conferences that are not about Abeta. This comfort zone for Abeta supporters has markedly limited clinical progress in the field of dementia research. It is important in science to unceremoniously bury a bad hypothesis, that is, one that is unsupported by clinical facts and one that simply does not work, as in the case of the Abeta concept. Not to do so is like keeping a score of old, smelly shoes in the closet that you are never going to wear.

References

1. Kuhn TS. The structure of scientific revolutions. 3rd ed. Chicago, Illinois: The University of Chicago Press; 1996. p. 24.
2. Hardy J, Selkoe DJ. The amyloid hypothesis of Alzheimer's disease: progress and problems on the road to therapeutics. Science. 2002;297:353–6.
3. Giannakopoulos P, Herrmann FR, Bussière T, Bouras C, Kövari E, Perl DP, Morrison JH, Gold G, Hof PR. Tangle and neuron numbers, but not amyloid load, predict cognitive status in Alzheimer's disease. Neurology. 2003;60(9):1495–500.
4. Davis DG, Schmitt FA, Wekstein DR, Markesbery WR. Alzheimer neuropathologic alterations in aged cognitively normal subjects. J Neuropathol Exp Neurol. 1999;58:376–88.

5. Braak H, Braak E. Frequency of stages of Alzheimerrelated lesions in different age categories. Neurobiol Aging. 1997;4:351–7.
6. Knopman DS, Parisi JE, Salviati A, Floriach-Robert M, Boeve BF, Ivnik RJ, et al. Neuropathology of cognitively normal elderly. J Neuropathol Exp Neurol. 2003;62:1087–95.
7. DeCarli C. Post-mortem regional neurofibrillary tangle densities, but not senile plaque densities are related to regional cerebral metabolic rates for glucose during life in Alzheimer disease patients. Neurodegeneration. 1992;1:11–20.
8. Irizarry MC, McNamara M, Fedorchak K, Hsiao K, Hyman BT. PPSw transgenic mice develop age-related A beta deposits and neuropil abnormalities, but no neuronal loss in CA1. J Neuropathol Exp Neurol. 1997;56:965–73.
9. Sassin I, Schultz C, Thal DR, Rüb U, Arai K, Braak E, Braak H. Evolution of Alzheimer's disease-related cytoskeletal changes in the basal nucleus of Meynert. Acta Neuropathol. 2000;100(3):259–69.
10. Liberski PP, Sikorska B, Hauw JJ, Kopp N. Ultrastructural characteristics (or evaluation) of Creutzfeldt-Jakob disease and other human transmissible spongiform encephalopathies or prion diseases. Ultrastruct Pathol. 2010;34:351–61.
11. Larson ME, Lesne SE. Soluble Abeta oligomer production and toxicity. J Neurochem. 2012;120(Suppl 1):125–39.
12. Benilova I, Karran E, De Strooper B. The toxic Aβ oligomer and Alzheimer's disease: an emperor in need of clothes. Nat Neurosci. 2012;15(3):349–57.
13. Robakis NK. Are Abeta and its derivatives causative agents or innocent bystanders in AD? Neurodeg Dis. 2010;7(1–3):32–7.
14. de la Torre JC. Phase 3 trials of solanezumab and bapineuzumab for Alzheimer's disease. N Engl J Med. 2014;370(15):1459–60.
15. Holmes C, Boche D, Wilkinson D, Yadegarfar G, Hopkins V, Bayer A, et al. Long-term effects of Abeta42 immunisation in Alzheimer's disease: follow-up of a randomised, placebo-controlled phase I trial. Lancet. 2008;372:216–23.
16. Salloway S, Sperling R, Fox NC, et al. Two phase 3 trials of bapineuzumab in mild-to-moderate Alzheimer's disease. N Engl J Med. 2014;370:322–33.
17. Doody RS, Thomas RG, Farlow M, et al. Phase 3 trials of solanezumab for mild-to-moderate Alzheimer's disease. N Engl J Med. 2014;370:311–21.
18. Kambhampaty A, Smith-Parker J. Eli Lilly's solanezumab faces grim prospects of attaining conditional FDA approval in mild Alzheimer's. Financial Times, 4 Sept 2012.
19. Ferrer I, Boada Rovira M, Sánchez Guerra ML, Rey MJ, Costa-Jussá F, et al. Neuropathology and pathogenesis of encephalitis following amyloid-beta immunization in Alzheimer's disease. Brain Pathol. 2004;14:11–20.
20. Mondragón-Rodríguez S, Basurto-Islas G, Lee HG, Perry G, Zhu X, Castellani RJ. Causes versus effects: the increasing complexities of Alzheimer's disease pathogenesis. Expert Rev Neurother. 2010;10:683–91.
21. Arriagada PV, Growdon JH, Hedley-Whyte ET, Hyman BT, et al. Neurofibrillary tangles but not senile plaques parallel duration and severity of Alzheimer's disease. Neurology. 1992;42:631–9.
22. Pathological correlates of late-onset dementia in a multicentre, community-based population in England and Wales. Neuropathology Group of the Medical Research Council Cognitive Function and Ageing Study (MRC CFAS). Lancet. 2001;357:169–75.
23. Rowe CC, Ellis KA, Rimajova M, Bourgeat P, Pike KE, Jones G, et al. Amyloid imaging results from the Australian Imaging, Biomarkers and Lifestyle (AIBL) study of aging. Neurobiol Aging. 2010;31:1275–83.
24. Keefe PR. The empire edge. The New Yorker, 13 Oct 2015. p. 76–89.
25. Shamoo AE, Resnick DB. Responsible conduct of research. 3rd ed. New York: Oxford University Press;2015. p. 106.

Chapter 6
Other Hypotheses on the Cause of Alzheimer's Disease

The Cholinergic Hypothesis

One of the most rewarding aspects of basic research is to discover the cause of a disease. The road to such a discovery requires investigators to display the patience of Buddhist monks and the ingenuity of Solomon during their research. Even then, there is no guarantee that the experiments will succeed. The reason is that one also needs luck. This seemed to be the case when t**he cholinergic hypothesis**, as it became known, was born in the mid-1970s following an observation made from Alzheimer's brain material. It was found that the enzyme choline acetyl transferase (CAT) was significantly reduced in the brains of people who had died of Alzheimer's disease [1].

CAT is the enzyme responsible for the formation of the important neurotransmitter acetylcholine, whose main function in the brain is the neuromodulation of learning, memory, and mood. The deduction from the finding of low CAT in Alzheimer's brain led to the obvious speculation about the role of acetylcholine in this dementia. Moreover, the CAT loss appeared most robust in hippocampal and neocortical regions, which are also favorite areas where early neurodegenerative changes begin to be seen in Alzheimer's [1].

The role of acetylcholine in Alzheimer's was strengthened when severe loss of cholinergic neurons in the basal nucleus of Meynert was found in Alzheimer's brains after death [2]. The basal nucleus of Meynert is a region that provides cholinergic input into the cortex and is, therefore, a primary neurotransmitter in cognitive function. Many researchers by now were somewhat hesitant to jump onto the acetylcholine bandwagon because it meant that Alzheimer's was a disease-specific disorder, which was too simple an explanation for such a complex dementia. A host of pharmaceuticals also decided to wait for more evidence before committing money to the cholinergic hypothesis. One crack in the cholinergic hypothesis was uncovered with the fact that many cognitively normal individuals tested also showed CAT deficiency in the brain. An astute psychiatrist, **Dr. William Summers**, became aware of the

J.C. de la Torre, *Alzheimer's Turning Point*,
DOI 10.1007/978-3-319-34057-9_6

clinical potential of the CAT loss in Alzheimer's brains and gave 12 patients oral tacrine, an enhancer of acetylcholine. Over several months of treatment, Summers reported measurable improvement of memory in half the patients using tacrine.

This finding resulted in a gold rush by pharmaceuticals to sort out their chemical shelves for cholinergic boosters and by another pharmaceutical to manufacture tacrine (tradename Cognex), by purchasing patents of the drug held by Summers. Although tacrine was eventually discontinued due to safety concerns, other cholinergic preparations were soon competing on the market not only in the USA but all over the world. These medications were similar to each other (so-called me-too drugs) and worked by blocking the enzyme, acetylcholinesterase, that breaks down acetylcholine at the synapse. This action raised the levels of acetylcholine in the brain and mildly improved cognitive impairment in Alzheimer's patients, especially at the mild stage. It was thought that the increase in acetylcholine entering the brain compensated for the loss of cholinergic neurons.

Data from our laboratory, however, indicated that the mildly beneficial effects of increasing acetylcholine in the brain with acetylcholinesterase inhibitors lay not necessarily in compensating for cholinergic neuron loss (since they were already dead) but in mildly increasing blood flow to the brain by the vasodilating effect of acetylcholine on cerebral vessels [3]. The fact that acetylcholinesterase inhibitors have only a very modest effect on Alzheimer's symptoms implies that the cholinergic increase of cerebral blood flow is marginal and that treatment with these drugs is given too late in the neurodegenerative process to reverse or block Alzheimer's progression [4].

Three cholinergic drugs are now the mainstay treatment for Alzheimer's. They are galantamine (Razadyne), rivastigmine (Exelon), and donepezil (Aricept). Although there is now a general agreement that cholinergic neuron loss is a secondary, not a primary, effect during Alzheimer's neuropathology, cholinergic drugs continue to hit pay dirt for the pharmaceutical industry in a big way. Each of these cholinesterase inhibitors brings more than a billion dollars to each drug company that sells them. This is a lucrative market that will get even better for the drug companies as the Alzheimer's population increases at an alarming rate.

A fourth drug on the market, memantine (Namenda), is not a cholinergic enhancer but works to suppress the activation of toxic products created during the neurodegenerative phase of Alzheimer's that can damage brain cells. These 4 drugs on the Alzheimer market are known to mildly and temporarily increase cerebral blood flow.

Ethical Considerations in Recommending Medication for Alzheimer's

What is the major action of these medications on Alzheimer's patients treated daily for long periods of time? Are the drugs of significant benefit to Alzheimer's patients with regard to symptoms, cognitive function, or daily quality of life? The response to these questions is tricky because drug companies are fond of engaging "publication planner" companies whose job is to ghostwrite medical articles for prominent doctors who are paid by the drug companies to praise the merits of their products.

The practice of skewing scientific information in a medical article by ghostwriters, who pose behind a prominent physician's name and the physician's affiliation with a distinguished university, is aided by prominent medical journals that turn a blind eye away from any impropriety. This conclusion has been argued by none other than the former editor of the distinguished New England Journal of Medicine, **Dr. Marcia Angell**, in her provocative book "The Truth About Drug Companies: How They Deceive Us and What To Do About It." This type of "quackademic medicine" is an important public health issue because patients may be prescribed medicine that is nor safe or medicines of dubious value. There is a lot of money riding on a favorable scientific article published in a respected medical journal that promotes the benefits of a drug product. This deplorable practice does not fall under the watchful eye of any authoritative regulatory body. With thousands of articles written every year on new drug product findings and clinical trials, it would be difficult to scrutinize every journal article for misleading information. However, the pharmaceutical industry should have the responsibility to be circumspect in its claims when presenting their product to the medical world. To exaggerate (or some would say "lie") about the effectiveness of a drug product or minimize its serious side effects is not only irresponsible and unprincipled but criminal, since physicians who read such reports rely on the veracity of the trial findings to prescribe the drug to patients. This is not always the case with pharmaceuticals that all-too-often have to withdraw a dangerous drug from the prescription market after many adverse events are reported by doctors, prompting the Food and Drug Administration (FDA) to force its removal as a prescriptive item. The FDA claims it does not have the staff to police the drug manufacturers into keeping a clean act.

One approach to maintain some objectivity in evaluating a drug product is to carefully and systematically review well-performed clinical trials that have assessed the potential benefit of a particular drug. This systematic review is called the **Cochrane Collaboration** which works by using meta-analysis of selected clinical trials. Meta-analysis is a statistical tool that aggregates results from different clinical studies in search of a clearer understanding of how well a treatment works. For example, a Cochrane analysis of the Alzheimer's treatment rivastigmine concluded:

"In comparisons with placebo, better outcomes were observed for rate of decline of cognitive function and activities of daily living, although the effects were small and of uncertain clinical importance." [5] The Cochrane review of donepezil was no less charitable. It concluded: **"There is little evidence that donepezil improved cognitive function, and no evidence exists that donepezil delays progression to Alzheimer's disease, but it was associated with significant side effects**." [6]

The high rate of adverse effects from cholinesterase inhibitors during the treatment of Alzheimer's is one of the reasons for many patients choosing to drop out from the treatment regimen. These side effects can be severe, especially at the initiation of treatment. They include nausea, vomiting, diarrhea, indigestion, and loss of weight. These side effects can result in moderate-to-severe dehydration leading to low blood pressure, rapid heartbeat, kidney failure and, in some cases, coma. Liver damage and liver failure have been reported in patients taking cholinesterase inhibitors for many months.

A cost-effective and systematic study of the literature was carried out in Great Britain on donepezil, rivastigmine, and galantamine in 2005 to analyze whether these drug treatments were clinically effective for mild to severe Alzheimer's [7]. This study found, to no one's surprise, that the benefits to Alzheimer's patients offered by these drugs were minimally modest as compared to a placebo and that reports from pharmaceutical sources that these medications display a positive effect on mood and behavior were "optimistic" and not substantiated by the evidence [7].

When Cholinesterase Inhibitors Are Not Prescribed

In Germany and China, cholinesterase inhibitors are prescribed much less frequently than the ancient herb ***Ginkgo biloba*** for patients with Alzheimer's disease and other dementias. The reasons are partly because ginkgo is considerably less costly than cholinesterase inhibitors and, according to many clinicians in Germany, better tolerated and just as effective as cholinesterase inhibitors. *Ginkgo biloba* has a long history in herbal medicine since it was mentioned in **Chinese Materia Medica** back in 2800 BC as a treatment of choice for vascular disorders, brain damaging conditions, and age-related ailments. Some common side effects from *Ginkgo biloba* therapy include constipation, upset stomach, rapid heartbeat, skin rash, and dizziness. As a blood thinner, gingko can modestly improve peripheral circulation and is reputed to increase memory function but it should not be taken with anticoagulants such as coumadin, aspirin, or the stroke preventive clopidogrel (Plavix) due to the risk of internal bleeding.

Other reasons for the success of *Ginkgo biloba* as a prescription drug for Alzheimer's in European countries is that it has been reported to increase brain blood flow and improve delivery of oxygen and glucose to the brain, the two fundamental energy precursors needed to fuel brain cells. Extracts of gingko also contain flavonoids and terpenoids, reputed to protect against mitochondrial damage and oxidative stress. Moreover, ginkgo has gained a reputation for improving social behavior, thinking, learning, and feeling of depression in Alzheimer's patients. A randomized controlled study on the benefits of *Ginkgo biloba* called Ginkgo Evaluation of Memory (GEM) was conducted in 2008 on 3000 elderly patients some with no Alzheimer's symptoms and others with mild cognitive impairment. Over the 6 years of evaluation, it was found that ginkgo was no different than placebo as far as ginkgo's effectiveness in reducing the incidence of Alzheimer's when preclinical mild symptoms were present [8]. Previous to the GEM study, it was reported that ginkgo does not provide any measurable benefit to memory decline in cognitively healthy adults who take oral ginkgo as a preventive medicine [9].

Memantine, the only Alzheimer's FDA-approved treatment on the market that does not target the cholinergic system, also did little to improve mental state or activities of daily life in Alzheimer's patients. It has a very modest positive effect on mood and behavior but has been reported to create problems for patients by increasing confusion, agitation, hallucinations, blood pressure, falls, and nervous system side effects [10].

Imagine if you are going to see a physician because someone in your family shows severe signs of memory impairment that have gotten progressively worse with time. The physician confirms your suspicions and you ask what treatments are available. The physician can take an hour or two to explain the pros and cons of memantine or cholinergic enhancers or the option of not treating with these drugs due to their side effects and marginal benefits. Or, the physician can whip out a prescription pad and write a prescription for one or two of the drugs discussed above. The latter option is the likely choice by most physicians. After all, time is money. Besides, the action fulfills two gratifying feelings for the physician. One, it shows the physician's willingness to help alleviate suffering despite recommending a medication of dubious value and serious side effects, and two, it does not waste the physician's time in explaining why these medicines are practically worthless. Besides, writing a prescription to assuage the pain of a grim diagnosis also lends support to the moronic rule that "something is better than nothing." The idea that nothing is better than something is too paradoxical to be a part of Western philosophy or culture.

The ethics of "to treat or not to treat" when moderate-to-severe Alzheimer's is diagnosed are rarely discussed in professional or lay articles and even non-profit organizations, such as the influential Alzheimer's Association in Chicago, who take no position at all on this matter in their advice to recently diagnosed Alzheimer's patients or their caregivers.

Cell Cycle Hypothesis

Cell cycle or cell division, in eukaryotic cells, is a highly regulated process involving growth and division of the cell into 4 distinct phases ending in mitosis and driven through the phases by enzymes called cyclin-dependent kinases. The biochemical processes involving cell cycle abnormalities are the objects of much study and have focused on terminally differentiated neurons which do not engage in detectable division but whose expression of cell cycle proteins is not silenced. Cell cycle-independent roles appear to be crucial to neuronal function in such activities as synaptic plasticity, learning, and memory [11]. It has been suggested, based on animal studies, that when a post-mitotic neuron reenters the cell cycle, it dies [12]. This phenomenon, if true, begs the question as to why reentry into the cell cycle by a post-mitotic neuron results in its death and under what circumstances does that happen?

The cell cycle hypothesis is based on the mitogenic cell cycle derangement observed in neurons from Alzheimer's brains [13]. Many cell cycle-related proteins appear to be associated with neurons that are vulnerable to neurodegeneration suggesting an early and pathologic role in the initiation of Alzheimer's dementia.

According to the cell cycle hypothesis, Alzheimer's-associated neurons progress from a resting phase (G_0) to the active (G_1) phase of the cell division and as they do so, a therapeutic target may present itself by preventing neurons from progressing to the S phase, preventing a gene dosage effect that can precipitate neurodegenerative changes.

Postmortem analysis of Alzheimer's brains suggests that cell cycle protein aberrations may precede the appearance of neurofibrillary tangles and could be responsible for their formation in Alzheimer's brain. In addition, evidence exists that cell cycle-induced neuronal death and the rate of cell loss is the same during the preclinical stage and all Alzheimer's stages that follow.

The hope is that since neurons engaged in cell cycle abnormalities will be metabolically whole compromised, interventions targeting this metabolic compromise may provide reversal of symptomatic in Alzheimer's victims. However, no evidence is available at the present that cell cycle interventions lower either Alzheimer's incidence or disease course.

Aluminum Hypothesis

The concept that aluminum could be the cause of Alzheimer's disease has been around for decades and may have begun in the 1920s when it was proposed that industrial exposure to aluminum resulted in memory failure [14]. **Dr. Donald Crapper McLachlan** and his colleagues [15] refined the aluminum hypothesis in 1973, when they reported that aluminum given to animals induced neurofibrillary degeneration in neurons. They also claimed that aluminum could be found in Alzheimer's brains at autopsy [15]. This concept was the source of hundreds of news articles in the lay press, TV reports, and medical conferences. People became concerned with exposure to aluminum foil, cooking pots, antiperspirants, and antacids containing aluminum salts.

There is little doubt that aluminum in high enough concentrations can be neurotoxic when chronically ingested or inhaled. But this is true for many other metals. An example of a universal toxic metal is mercury, found in dental amalgams and fish. Lead, arsenic, and cadmium found in foods, drinking water, pesticides, and cigarettes are other common metallic compounds that can cause an assortment of illnesses, including heart disease, cancer, and brain damage.

Several fundamental problems have plagued the aluminum hypothesis. For example, its exposure varies from country to country, whereas Alzheimer's disease incidence is rather constant throughout the world where such records are kept. This fact indicates that while heavy exposure of aluminum can cause neurodegenerative brain damage affecting memory, it is not the cause of Alzheimer's dementia, since countries with low aluminum exposure have the same incidence of Alzheimer's as countries where aluminum exposure is common.

Another inconsistency with the aluminum hypothesis is the finding that there is no significant difference in aluminum brain levels in Alzheimer's patients when compared to age-matched controls [16]. Moreover, the effects of long-term exposure to aluminum resulting in chronic renal failure and death were studied in the brains of patients postmortem. It was found that no patient showed the formation of neurofibrillary tangles or senile plaques, the hallmark markers of Alzheimer's [17].

Finally, several epidemiological studies have failed to find a relationship between aluminum toxicity and Alzheimer's disease [18, 19]. Evidence suggests that aluminum neurotoxicity could be damaging not only to elderly brain but also to young brain, but this damage does not appear to promote the neuropathology that is consistent with Alzheimer's. It is, therefore, unlikely that aluminum plays a pathogenic role in Alzheimer's.

Inflammatory Hypothesis

It is unusual for a neurological disorder not to show some degree of inflammatory activity which can be beneficial or, in many cases, harmful. In the case of Alzheimer's, despite an absence of inflammatory leukocyte influx into the brain, there appears to be a wake-up call of microglia, the cells who act as anti-inflammatory watch dogs in the brain and function very similarly as macrophages do outside the brain. Although the brain is considered to be immunologically privileged and outside of immune control, it is not defenseless since cerebral microglia can raise the alarm when it detects intruders and, if necessary, attack those using inflammatory responders or mediators. As resident phagocytes in the central nervous system, microglia can mobilize the release of molecules that can promote inflammation which can either combat or worsen neurodegeneration. These proinflammatory molecules include free radicals, chemokines, proteases, and other inflammation-related responders. This is a normal reaction because if injured tissue, including brain, did not show signs of inflammation during or after degenerative changes, the outcome would be considered abnormal [20]. Benefits of inducing neuroinflammation in the brain include neuroprotection, that is, the mobilization of neural precursors for repair, remyelination, and even axonal regeneration following damage to neural tissue [20].

The inflammatory hypothesis maintains that inflammation significantly contributes to the development of Alzheimer's pathology [21]. Supporters of this idea speculate that inflammation in Alzheimer's dementia may be due to the presence of amyloid plaques which contain the major complement proteins that are normally associated with immune activation [21]. In Alzheimer's brain, activated microglia have been reported to cluster around senile plaques while producing massive amounts of lethal oxygen radicals and inflammatory molecules. These mediators are toxic to brain cells ultimately destroying them [22].

There is substantial evidence in animal models and human Alzheimer's showing that a sustained inflammatory response involving microglia and astrocytes contribute to disease progression and the destruction of neural tissue [22]. For this reason, many investigators have suggested targeting inflammatory reactions for slowing or delaying the onset of Alzheimer's progression by using anti-inflammatory agents that ostensibly contribute to blocking inflammation.

The best supporting evidence for a causal role of inflammation in Alzheimer's is contained in several epidemiological studies supporting the role of inflammation in dementia. One study concluded that inflammatory proteins are increased prior to the clinical onset of Alzheimer's disease [23]. Although this finding was of interest, it did not necessarily prove that inflammatory proteins trigger neurodegeneration. These epidemiological studies came into question when more rigorous randomized trials were performed using placebo-controlled versus anti-inflammatory drugs such as ibuprofen, celecoxib, and naproxen. The anti-inflammatory drugs failed to show any benefit in reducing early cognitive impairment of Alzheimer's disease [24, 25].

For the last 25 years, dozens of observational studies have examined the relationship between taking nonsteroidal anti-inflammatory drugs (NSAIDs) and the risk of contracting Alzheimer's disease in humans. The data from these studies are controversial with findings ranging from no effect in modifying Alzheimer's risk when using NSAIDs to reducing the risk of Alzheimer's prior to disease onset. However, because observational studies pose a potential or overt bias, randomized controlled trials were more recently conducted to examine the effects of NSAIDs in lowering Alzheimer's risk. A number of different anti-inflammatory agents including indomethacin, diclofenac, nimesulide, naproxen, celecoxib, rofecoxib, and ibuprofen were clinically tested in randomized controlled trials by different independent groups. The results were largely disappointing and generally negative [26]. No slowdown or reversal of cognitive impairment was observed when anti-inflammatory agents were used as treatments. [26, 27], Although most people tolerate NSAIDs without any difficulty, side effects can occur, including kidney and liver toxicity, increased hypertension, and gastrointestinal damage.

At the present, the controversy continues about whether inflammation is the primary trigger or a consequence of other pathologic events occurring preclinically in people with mild memory impairment. However, inflammation is classically considered a response triggered by damage to living tissues. Many workers in the field lean toward the concept that inflammation does not represent a causal factor in neurodegenerative diseases, including Alzheimer's. Inflammation is more likely a downstream event occurring during the process of neurodegeneration.

Since anti-inflammatory agents presently on the market do not seem to help ward off or benefit those at risk of dementia, they cannot be recommended at this time for that purpose. In the event that more selective anti-inflammatory drugs are developed in the future that can indeed impact Alzheimer's onset or cognitive loss, NSAIDs should be left in the medicine cabinet and used for other indications that do not involve cognitive behavior.

Oxidative Stress Hypothesis

Oxidative stress is a damaging condition that occurs when oxidant production in mammalian cells releases lethal quantities of reactive oxygen species (called ROS) such as toxic-free radicals that eventually overwhelm antioxidant reserves held by the cell. There is a consensus that the overproduction of toxic-free radicals generated from oxidative stress has a major role in neurodegeneration [28] and, as a damaging reactive process, it may be involved in cell cycle abnormalities contributing to cell death and organ damage [29]. Oxidative stress never acts alone or without the accompanying oxidation–reduction (redox), a chemical reaction involved in cell energy production and generally triggered by stress molecules.

In brain, a variety of stress molecules can induce oxidative stress. These stress moles (or stressants) include low blood flow to the brain, inflammation, aging, lower oxygen in brain, cigarette smoking, excess alcohol drinking, cardiovascular disease, and a host of other conditions [30]. Many of these metabolic insults can contribute to Alzheimer's pathology but are not disease-specific since they are observed in other neurodegenerative disorders [31].

The oxidative stress hypothesis is based on the finding that in the Alzheimer's brain, increased trace metals such as iron, copper, and mercury occur in the presence of increased lipid peroxidation, where lipids are degraded by toxic-free radicals and become a crucial step in the development of DNA breakage and tissue damage. Lipid peroxidation occurs in disease states such as Alzheimer's disease, heart failure, rheumatic arthritis, cancer, and assorted immunological disorders [32]. Once the process of lipid peroxidation begins, mitochondrial metabolism decreases which can affect the production of energy nutrients needed for brain cell survival. It is believed that the chronologic appearance of these changes involving oxidative stress occurs prior to Alzheimer's and during the evolution of this dementia [33].

The Calcium Hypothesis

Calcium (Ca^{2+}) is a key molecule in cells of the body. Once Ca^{2+} enters a cell membrane, it can influence enzymes and proteins inside the cell by a series of complex reactions. The movement of Ca^{2+} from the outside to the inside of a neuronal cell creates a brief electrical impulse on the cell membrane called an action potential. The action potential delivers a nerve impulse to other neurons to regulate neurotransmission at the synapse.

The calcium hypothesis attempts to explain how age-related neuronal Ca^{2+} dysfunction leads to amyloid formation and accumulation of Abeta in the brain to generate memory loss and cell death in Alzheimer's brain. The abnormal metabolic pathways involved in Ca^{2+} dysfunction appear to occur long before the neurodegenerative or cognitive Alzheimer's changes are detected [34]. Thus, Ca^{2+} disruption would be the principal trigger to the formation of Abeta-containing plaques and neurofibrillary tangles.

Some supporters of the calcium hypothesis see a **bidirectional relationship between Ca^{2+} signaling and Abeta accumulation in Alzheimer's brain** [35]. Calcium signaling is a system that regulates numerous biological processes, especially in excitable cells (neurons, muscle cells) and in tissue physiology. In this scenario, disruption of calcium signaling induces neurometabolic abnormalities and excess production of Abeta leading to Alzheimer's. Unfortunately, no study, so far, has provided solid evidence that **Ca^{2+}** dysregulation in neurons causes Abeta production or the other way around, that Abeta production causes Ca^{2+} dysregulation.

If **Ca^{2+}** signaling disturbances could be confirmed preceding Alzheimer's pathology, an effective treatment could be designed targeting intracellular **Ca^{2+}** that might prevent further neuronal damage and cognitive decline. Treatments to reduce intracellular Ca^{2+} concentrations in brain, such as lithium, have been tried without success. Neither has reducing Abeta in the brain using immunotherapy worked as was previously discussed under the Abeta hypothesis. This fact does not add support to a bidirectional pathologic pathway between **Ca^{2+}** dysfunction and Abeta in propelling the start of dementia.

Despite intense research into the possible causes of Alzheimer's disease, no clinical treatment has been found that alters the disease course in any way. This is a frustrating and costly exercise because without revealing the nature of how Alzheimer's begins, the treatment or cure remains elusive (Fig. 2.1). This has led to an intense search in many research laboratories to find a winning permutation that can crack open the conundrum surrounding this dementia. So far, few investigators have asked the crucial question "are we on the right track?"

References

1. Davies P. Challenging the cholinergic hypothesis in Alzheimer disease. JAMA. 1999;281(15):1433–4.
2. Whitehouse PJ, Price DL, Struble RG, Clark AW, Coyle JT, Delon MR. Alzheimer's disease and senile dementia: loss of neurons in the basal forebrain. Science. 1982;215(4537):1237–9.
3. de la Torre JC. Hemodynamic consequences of deformed microvessels in the brain in Alzheimer's disease. Ann NY Acad Sci. 1997;26(826):75–91.
4. de la Torre JC. Alzheimer disease as a vascular disorder: nosological evidence. Stroke. 2002;33(4):1152–62.
5. Birks J, Flicker L. Donepezil for mild cognitive impairment. Cochrane Database Syst Rev. 2006;(3):CD006104.
6. Birks JS, Grimley Evans J. Rivastigmine for Alzheimer's disease. Cochrane Database Syst Rev. 2015;4:CD001191.
7. Loveman E, Green C, Kirby J, Takeda A. The clinical and cost-effectiveness of donepezil, rivastigmine, galantamine and memantine for Alzheimer's disease. Health Technol Assess 2006;10:iii–iv, ix–xi, 1–160.
8. DeKosky ST, Williamson JD, Fitzpatrick AL. *Ginkgo biloba* for prevention of dementia. J Am Med Assoc. 2008;300(19):2253–62.
9. Solomon PR, Adams F, Silver A, Zimmer J, DeVeaux R. Ginkgo for memory enhancement: a randomized controlled trial. JAMA. 2002;288(7):835–40.

10. Yang Z, Zhou X, Zhang Q. Effectiveness and safety of memantine treatment for Alzheimer's disease. J Alzheimers Dis. 2013;36(3):445–58.
11. Angelo M, Plattner F, Giese KP. Cyclin-dependent kinase 5 in synaptic plasticity, learning and memory. J Neurochem. 2006;99:353–70.
12. Herrup K, Busser JC. The induction of multiple cell cycle events precedes target-related neuronal death. Development. 1995;121:2385–95.
13. Bowser R, Smith MA. Cell cycle proteins in Alzheimer's disease: plenty of wheels but no cycle. J. Alzheimer's Dis. 2002;4:249–54.
14. Shore D, Wyatt RJ. Aluminum and Alzheimer's disease. J Nerv Mental Dis. 1983;171 (9):553–8.
15. Crapper DR, Krishnan SS, Dalton AJ. Brain aluminum distribution in Alzheimer's disease and experimental neurofibrillary degeneration. Science. 1973;180(4085):511–3.
16. Bjertness E, Candy JM, Torvik A. Content of brain aluminum is not elevated in Alzheimer disease. Alzheimer Dis Assoc Disord. 1996;10(3):171–4.
17. Candy JM, McArthur FK, Oakley AE. Aluminium accumulation in relation to senile plaque and neurofibrillary tangle formation in the brains of patients with renal failure. J Neurol Sci. 1992;107(2):210–8.
18. Wettstein A, Aeppli J, Gautschi K, Peters M. Failure to find a relationship between mnestic skills of octogenarians and aluminum in drinking water. Int Arch Occup Environ Health. 1991;63(2):97–103.
19. Martyn CN, Coggon DN, Inskip H, Lacey RF, Young WF. Aluminum concentrations in drinking water and risk of Alzheimer's disease. Epidemiology. 1997;8(3):281–6.
20. WeeYongV. Inflammation in neurological disorders: a help or a hindrance? Neuroscientist. 2010;16:408–20.
21. McGeer PL, McGeer EG, Yasojima K. Alzheimer disease and neuroinflammation. J Neural Transm. 2000;Suppl 59:53–7.
22. Glass CK, Saijo K, Winner B, Marchetto MC, Gage FH. Mechanisms underlying inflammation in neurodegeneration. Cell. 2010;140:918–34.
23. Engelhart MJ, Geerlings MI, Meijer J, Kiliaan A, Ruitenberg A, van Swieten JC, Stijnen T, Hofman A, Witteman JC, Breteler MM. Inflammatory proteins in plasma and the risk of dementia: the Rotterdam Study. Arch Neurol. 2004;61:668–72.
24. Breitner JC, Haneuse SJ, Walker R, Dublin S, Crane PK, Gray SL, Larson EB. Risk of dementia and AD with prior exposure to NSAIDs in an elderly community-based cohort. Neurology. 2009;72:1899–905.
25. Aisen PS, Schafer KA, Grundman M, Pfeiffer E, Sano M, Davis KL, Farlow MR, Jin S, Thomas RG. Effects of rofecoxib or naproxen vs placebo on Alzheimer disease progression: A randomized controlled trial. JAMA. 2003;289:2819–26.
26. Szekely CA, Zandi PP. Non-steroidal anti-inflammatory drugs and Alzheimer's disease: the epidemiological evidence. CNS Neurol Disord Drug Targets. 2010;9(2):132–9.
27. Aisen PS. Alzheimer's disease therapeutic research: the path forward. Alzheimers Res Ther. 2009;1(1):2.
28. Harman D, Eddy DE, Noffsinger J. Free radical theory of aging: inhibition of amyloidosis in mice by antioxidants; possible mechanisms. J Am Geriatr Soc. 1976;24:203–10.
29. Yang Y, Mufson EJ, Herrup K. Neuronal cell death is preceded by cell cycle events at all stages of Alzheimer's disease. J Neurosci. 2003;23:2557–63.
30. Guglielmotto M, Tamagno E, Danni O. Oxidative stress and hypoxia contribute to Alzheimer's disease pathogenesis: two sides of the same coin. Sci World J. 2009;9:781–91.
31. Kamat CD, Gadal S, Mhatre M, Williamson KS, Pye QN, Hensley K. Antioxidants in central nervous system diseases: preclinical promise and translational challenges. J Alzheimers Dis. 2008;15:473–93.
32. Ramana KV, Srivastava S, Singhal S. Lipid peroxidation products in human health and disease. Oxidative Med Cell Longevity. 2013;2013, Article ID 583438.

33. Smith MA, Zhu X, Tabaton M, Liu G, McKeel DW Jr, Cohen ML, Wang X, Siedlak SL, Dwyer BE, Hayashi T, Nakamura M, Nunomura A, Perry G. Increased iron and free radical generation in preclinical Alzheimer disease and mild cognitive impairment. J Alzheimers Dis. 2010;19:363–72.
34. Stutzmann GE. The pathogenesis of Alzheimer's disease—is it a lifelong "calciumopathy". Neuroscientist. 2007;13:546–59.
35. Green KN, LaFerla FM. Linking calcium to Aβ and Alzheimer's disease. Neuron. 2008;59:190–4.

Chapter 7
Alzheimer's Noise

When Hope Becomes Hype

This is a time when newspapers, magazines, and TV pundits jump at the slightest chance of reporting Alzheimer's-related stories that can capture national attention. This "haste without debate" often leads to a daily scattering of unfounded "breakthroughs," "better understandings," and "discoveries" that are never heard of again. Most of the attention-grabbing news coverage on Alzheimer's continues to dispense false hope, confusion, and wildly misleading hype by featuring goofy medical claims that offer little or zero understanding of the complexity involved in this disease. Some claims border on the absurd, like the domestic fly serving as an Alzheimer's model.

Unproven treatments for Alzheimer's are often prattled by the press to the public without the slightest consideration for factual evidence or thought-provoking viewpoints. If expert opinion is sought by the media, the same experts are chosen year after year who cherry-pick their favorite personal bias of any arguable issue. These experts are often chosen on the recommendation of their institution's public relations office or their pharmaceutical connections, not necessarily for any significant contribution made to science.

Brain games are often advertised as "brain training" that can prevent Alzheimer's, but there is no scientific evidence to back up any of these claims. The Federal Trade Commission (FTC) recently levied a $2 million fine to Lumos Labs, a company that advertised its brain game Lumosity as able to boost performance at work or school and delay the onset of Alzheimer's. These unproven claims are not only fraudulent but they prey on consumer gullibility and fears to sell their products. False advertising of brain-enhancing applications is now a billion-dollar industry.

What is the public to think when they are bombarded daily with a litany of panaceas ranging from coconut oil and aluminum chelation to cactus juice and mustard greens as remedies for Alzheimer's symptoms?

J.C. de la Torre, *Alzheimer's Turning Point*,
DOI 10.1007/978-3-319-34057-9_7

A commonplace example of how media critics and TV news presented a test that claims to predict who will get Alzheimer's and who will not is worth mentioning. This test relies on using a PET scan to visualize amyloid beta (Abeta) in the brain. Abeta is a protein associated with Alzheimer's histopathology but it can also be found in the brains of cognitively healthy people who never develop the disease [1, 2]. Aside from being wildly expensive, ($6000 per scan) the test to detect Abeta, known as PiB, underwent meta-analysis (a careful assessment of the clinical evidence) derived from a collection of previous studies to determine its diagnostic usefulness. Meta-analysis indicated the PIB test is imprecise and did not recommend its routine use in clinical practice [3].

Moreover, the Medicare Evidence Development and Coverage Advisory Committee (MEDCAC) expressed skepticism of the PIB (Amyvid) test. It concluded in 2013 that Amyvid and similar agents under development could not reliably diagnose Alzheimer's or the risk of developing it by the presence of Abeta in a brain scan [4]. The MEDCAC panel recommended that no reimbursement by Medicare or Medicaid be allowed for the Amyvid test in patients with suspected dyscognitive symptoms. The MEDCAC panel pointed out that since 30 % of cognitively healthy, elderly people have Abeta deposition in their brains, widespread testing has the potential to do more harm than good.

Little thought seems to have been given to the consequences of testing for Abeta deposits in the brain with PIB. First, it must be terrifying to be told that Alzheimer's is going to strike you soon, especially since the test is unreliable and you may never develop the disease, and second, because even if you do contract Alzheimer's, there is no cure or therapy available that will change the course of the disease.

Dr. Kenneth Covinsky, a professor of geriatrics at the University of California, San Francisco, summarized this dilemma:

> The challenge here is in distinguishing risk for disease with disease itself. There is considerable evidence that many, if not most people with these [brain] biomarkers, will never get Alzheimer's disease. For example, autopsy finding suggestive of Alzheimer's disease are commonly found in people who never had symptoms [5]

Adding to this caution, it is hard not to view these "predictive" Alzheimer's tests with a cynical eye considering the money they generate for the promoters, their institutions, and the sponsoring pharmaceuticals.

As a counter to this mindless practice, Chapter 15 will examine a practical approach, free of pharmaceutical influence, to identify, detect, and intervene in cases where impending cognitive deterioration is discovered in the asymptomatic or mildly symptomatic person.

During the last decade, after repeated clinical trial failures of anti-amyloid therapies, many investigators realized that continuing to work on the Abeta hypothesis was like beating a dead lizard and decided to switch their attention to a different research facet of Alzheimer's dementia. These workers found the vascular hypothesis of Alzheimer's disease to be a reasonable alternative in explaining much of the reported Alzheimer's pathology.

Eliminating Alzheimer's Disease

The vascular hypothesis of Alzheimer's disease now has a considerable following among investigators, and its growing research spectrum has become a mother-lode of important advancements in managing and treating this disorder (see reviews [6]–[11]).

This can be the **turning point in the fight to thwart Alzheimer's onset** because now, for the first time, there is a realistic clinical target where potential interventions can be applied to substantially prevent this insidious disorder. For this reason, the journey to prevent Alzheimer's dementia now seems to be on the cusp of being realized probably more so than in the last 100 years since it was first described by the German neuropathologist Alois Alzheimer [12].

How soon will Alzheimer's be eliminated as the unavoidable curse of aging? That depends to a large extent on the consolidated interest of scientists to think outside the box and on the research funding provided by a fully committed government program. No small task but definitely achievable, as reviewed in the pages ahead and in Chap. 16.

Much has been written on the subject of warding-off Alzheimer's and some of the advice is sound and legitimate but other suggestions can be distracting, irrelevant, or unsupported by medical evidence. Most Alzheimer's books on the market are either about trite ways to dodge Alzheimer's or guides aimed at the caretakers. The caretaker guide books, in a sense, are a useful gathering of ideas to relieve the tremendous pressure of caring for someone who may be rapidly progressing into a cognitive collapse. A few books deal with testimonials from actual Alzheimer's patients at the early stages of the disease.

The 2014 movie *Still Alice* is based on a novel by neuroscientist **Lisa Genova** that relates to a fictitious 50-year-old linguistics professor faced with early-onset Alzheimer's and her swift mental decline from a quick-witted academic to an empty shell of herself. Although early-onset Alzheimer's dementia is an extremely rare version of the common, non-genetic Alzheimer's type, the book offers insight into the progressive deterioration of the mind and body and the consideration of suicide by the heroine who realizes the indignity of gradually losing control of her intellect and behavior. Suicide in the context of Alzheimer's disease is discussed in Chap. 11, under **Ugly aging**. One of the main points made in *Still Alice* is that behind every Alzheimer's patient, there is a family that is torn apart by the tragedy of losing a loved person whose mind slowly drifts away with each passing day.

Other movies such as *Iris*, based on **Iris Murdoch's** struggle with Alzheimer's and *The Iron Lady*, a real-life story about British Prime Minister **Margaret Thatcher's** battle with dementia, point to the public's growing concern and interest in learning more about this brain-wasting and seemingly unstoppable disease. Despite the public's present interest in Alzheimer's, it was not always like that.

Consider that in 1974, the classic 1,600 page textbook *Pathologic Basis of Disease* by **Dr. Stanley Robbins** [13], devoted barely a page to describe Alzheimer's disease. In that sense, one could say that Alzheimer's is a relatively new brain disorder recognized in the last 4 decades as a condition that slowly takes the brain for a lethal ride to a cryptic point in space where there is no return. The awareness of Alzheimer's devastating impact was probably known by doctors since the turn of the twentieth century but no one imagined then it would one day affect 40 % of the people from age 75–84.

President Reagan Develops Alzheimer's Disease—Was Poor Brain Blood Flow the Trigger?

On March 30, 1981, three months after being elected, **President Ronald Reagan** was shot by John Hinckley Jr. as he was leaving the Washington Hilton Hotel in the nation's capital. The 22-caliber bullet ricocheted off the side of Reagan's limousine and struck the president on the left underarm where it lodged into his lung causing profuse bleeding. At the hospital, the president's left lung was found collapsed and his systolic blood pressure dropped to 60, an ominous sign indicating that shock and hypoxemia was present. A further dip in blood pressure can quickly turn into a death spiral. Taken to the operating room, the president was seen to lose half his blood volume during the 90-min surgical exploration to find and remove the bullet from his lung. This amount of blood loss can quickly lead to hemorrhagic shock, a condition featuring reduced oxygen delivery, tissue hypoperfusion, hemodynamic instability, cell hypoxia, and the threat of death. The treatment is to stop the bleeding and restore circulating blood volume.

According to hospital records, President Reagan first received 900, then 1200, and then 1800 cc of blood due to his massive hemorrhaging during his ordeal. His surgeon recalled that he had never seen such loss of blood in similar lung injuries and ventured that if the bleeding had not been controlled for another 5 min, the president would have reached a point of no return where death or permanent brain injury was inevitable.

At age 70, Ronald Reagan was the oldest president in US history. At that age, cerebral blood flow normally drops about 20–25 % from age 20. This is not a problem with respect to cognitive impairment unless another burden is added to the 20 % cerebral blood flow drop. This burden can occur with sudden and significant loss of blood volume or to the presence of vascular risk factors. The presence of several vascular risk factors to Alzheimer's (see Figs. 12.1, 12.2 and 12.3) after age 60 is now known to add an important burden to brain blood flow. In Reagan's case, no vascular risk factors to Alzheimer's were made known to the public except for his family history (his mother died of dementia), and smoking, which the president had given up years before the assassination attempt.

The president was severely hypoxic for many hours following his admission to the hospital, and this is evident from his collapsed left lung and his complaint of breathing difficulties. The president was likely brain ischemic by reason of his massive blood loss which was reflected by his low systolic blood pressure upon admission. This ischemic state is also suggested by his inability to walk properly when first admitted, his feeling lightheaded and his disorientation when asked simple questions by the attending doctors and nurses. No record is available that his cerebral blood flow was ever measured at any time during his hospital stay or during his recovery. The duration of brain ischemia was also not recorded or reported.

After his recovery, President Reagan returned to the White House and carried on his presidential duties until the end of his term in office in 1990. His son Ron Reagan noted in his book *My Father at 100* that his dad showed signs of dementia while in office. Ron Reagan was in his mid 20s at the time his father was president. The observation by the president's son is supported by a study recently published from the University of Arizona that examined 46 news conferences held by President Reagan during his two terms in office. This evidence, if one can call it that, points to President Reagan's use of vague nouns like "something" or "anything" to replace more complex words. These speech changes can be signs of mild cognitive impairment and impending Alzheimer's. Moreover, the president showed a penchant for contradictory statements, absentmindedness, and forgetting names of close friends or places which was uncharacteristic of his behavior prior to his assassination attempt. It was a well-guarded observation that the president's wife Nancy stood at times behind the president whispering to him some words to utter when the president went blank after being asked questions from reporters. The president admitted he had acquired Alzheimer's in 1994, 3 years after leaving office but never disclosed when his cognitive decline began to appear which can be assumed started many years before.

It is now known that mild hypoxemia from sleep apnea and chronic obstructive pulmonary disorder (COPD) is associated with cognitive deficits due to inadequate oxygen reaching organ cells in the body, including the brain. Mild hypoxia can affect memory, attention, decision-making, and use of words [14].

These cognitive deficits were frequently observed while President Reagan was in office, and **it is tempting to speculate that many socioeconomic decisions affecting world events during President Reagan's term in office were tainted by flawed thinking**.

It can be argued that it is difficult to prove an acute episode of severe blood loss and hypoxia at age 70 such as President Reagan sustained can trigger progressive cognitive decline and later Alzheimer's. There is no longitudinal study to confirm or negate such an association. However, a sudden acute episode of ischemic stroke can induce permanent mild cognitive deficits even when the stroke is successfully treated several hours after onset [15]. Moreover, the fact that chronic brain hypoperfusion may precede symptoms of severe cognitive impairment by a decade or more seems to correlate with President Reagan's slow evolution of Alzheimer's pathology. Although there is no official evidence that President Reagan was neurologically or cognitively evaluated after his surgical recovery, it is reasonable to assume that minor damage to his brain may have occurred during a long period of hypoxia which likely affected his cognitive function. This has been shown in patients who recover fully after a transient ischemic attack where mild cognitive fragility continues well beyond the resolution of focal neurological symptoms [16].

For those reasons, it is reasonable to assume that President Reagan's bout with brain hypoperfusion for several hours after being shot, especially at his advanced age, could have resulted in a short but clinically significant period of global hypoxic–ischemia sufficient to unleash a neuronal energy crisis. A neuronal energy crisis will involve neuronal damage that is often quickly lethal (as in stroke) but can also develop very slowly, as in chronic brain hypoperfusion. This energy crisis can also extend to astrocytes that control vessel blood flow in the brain (Fig. 7.1) (see our review [17]. The main element that characterizes a neuronal energy crisis after an acute episode of brain ischemia is an inadequate supply of blood and oxygen delivered to the brain. The resulting hypoxemia can generate mitochondrial damage and sudden loss of the cell energy fuel ATP leading to minimal or extensive brain cell death [18] (Fig. 7.1).

Death of neurons after a neuronal energy crisis does not necessarily appear immediately but may be manifested weeks or months after the initial insult. The pattern of slow cognitive decline following a neuronal energy crisis is

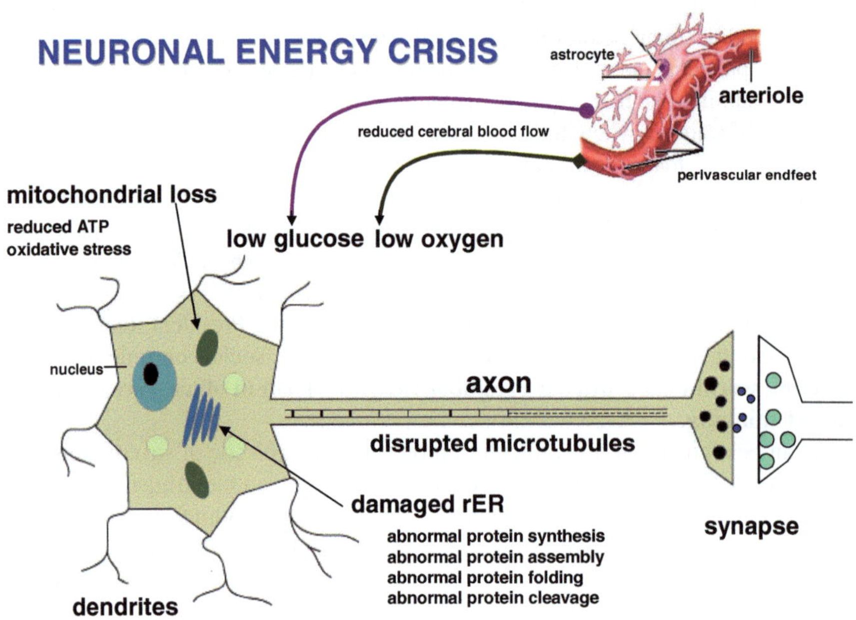

Fig. 7.1 Simplified theoretical sketch involved in a neuronal energy crisis following acute or chronic cerebral blood flow insufficiency. Brain hypoperfusion can result quickly from an acute ischemic stroke or slowly from chronic low blood flow delivery to brain cells. Diminished delivery of glucose and oxygen can result in a reduction of the energy fuel ATP from mitochondrial loss or damage. An energy crisis within the neuron will act with a domino effect to induce oxidative stress, protein synthesis abnormalities from damage the rough endoplasmic reticulum (*rER*), synaptic dysfunction, and neurotransmitter impairment from disrupted microtubules. These changes can impair neuronal function or result in neuronal death in ischemic-sensitive brain regions such as the hippocampus, a center for memory and learning

compatible with President Reagan's progressive mental slips and memory lapses while in office. Moreover, his advanced age made it more difficult to cushion the progressive neuronal pathology stemming from an unrelenting brain hypoperfusion that probably began after being shot.

It seems ironic that had the president's wife Nancy taken up the cause of the pitiful research funding for Alzheimer's during her husband's struggle with the disease, we might know more today about this dementia and the best ways to prevent it. This lost opportunity might also have helped tough gun control regulation since gunshot wounds to the head will devastate cognitive function in most survivors. This is evidenced by James Brady, the president's press secretary who was shot in the head during the president's assassination attempt and who suffered memory and thinking deficits until his death in 2014. Although the Brady Bill of handgun control was enacted in 1993, it has had no discernable impact on reducing gun-related deaths in the U.S.

References

1. Giannakopoulos P, Herrmann FR, Bussière T, Bouras C, Kövari E, Perl DP, Morrison JH, Gold G, Hof PR. Tangle and neuron numbers, but not amyloid load, predict cognitive status in Alzheimer's disease. Neurology. 2003;60(9):1495–500.
2. Rodrigue KM, Kennedy KM. Park DC Beta-amyloid deposition and the aging brain. Neuropsychol Rev. 2009;19:436–50.
3. Zhang S, Smailagic N, Hyde C, Noel-Storr AH, Takwoingi Y, McShane R, Feng *(11)C*-PIB-PET for the early diagnosis of Alzheimer's disease dementia and other dementias in people with mild cognitive impairment (MCI). J. Cochrane Database Syst Rev. 2014 Jul 23;7: CD010386.
4. Pierson R. U.S. Medicare panel skeptical of brain-plaque scans. Lifescropt. January 31, 2013.
5. Covinsky K. Caution on diagnosing preclinical Alzheimer's disease. Bioethics Forum. 2011; July 15.
6. Meng XF, Yu JT, Wang HF, Tan MS, Wang C, Tan CC, Tan L. Midlife **vascular** risk factors and the risk of Alzheimer's disease: a systematic review and meta-analysis. J Alzheimers Dis. 2014;42(4):1295–310.
7. Ruitenberg A, den Heijer T, Bakker SL, van Swieten JC, Koudstaal PJ, Hofman A, Breteler MM. Cerebral hypoperfusion and clinical onset of dementia: the Rotterdam study. Ann Neurol. 2005;57(6):789–94.
8. Farkas E, Luiten PG. Cerebral microvascular pathology in aging and Alzheimer's disease. Prog Neurobiol. 2001;64:575–611.
9. Mazza M, Marano G, Traversi G, Bria P, Mazza S. Primary cerebral blood flow deficiency and Alzheimer's disease: shadows and lights. J Alzheimers Dis. 2011;23(3):375–89.
10. Johnson NA, Jahng GH, Weiner MW, Miller BL, Chui HC, Jagust WJ, Gorno-Tempini ML, Schuff N. Pattern of cerebral hypoperfusion in Alzheimer disease and mild cognitive impairment measured with arterial spin-labeling MR imaging: initial experience. Radiology. 2005;234(3):851–9.
11. Sabayan B, Jansen S, Oleksik AM, van Osch MJ, van Buchem MA, van Vliet P, de Craen AJ, Westendorp RG. Cerebrovascular hemodynamics in Alzheimer's disease and vascular dementia: a meta-analysis of transcranial Doppler studies. Ageing Res Rev. 2012;11 (2):271–7.

12. Stelzma RA, Schinitzlein H, Muriagh FR. An English I'ranslation of Alzheimer's 1907 Paper, " Uber eine eigenartige Erkrankung der Hirnrinde". Clin Anat. 1995;8:429–31.
13. Robbins SL. Pathologic basis of disease. Philadelphia: W.B. Saunders Company; 1974.
14. Pighin S, Bonini N, Savadori L, Hadjichristidis C, Schena F. Loss aversion and hypoxia: less loss aversion in oxygen-depleted environment. Stress. 2014;17(2):204–10.
15. Sun JH, Tan L, Yu JT. Post-stroke cognitive impairment: epidemiology, mechanisms and management. Ann Transl Med. 2014;2(8):80.
16. Pendlebury ST, Wadling S, Silver LE, Mehta Z, Rothwell PM. Transient cognitive impairment in TIA and minor stroke. Stroke. 2011;42(11):3116–21.
17. de la Torre JC. Pathophysiology of neuronal energy crisis in Alzheimer's disease. Neurodegener Dis. 2008;5(3–4):126–32.
18. Weilinger NL, Maslieieva V, Bialecki J, Sridharan SS, Tang PL, Thompson RJ. Ionotropic receptors and ion channels in ischemic neuronal death and dysfunction. Acta Pharmacol Sin. 2013;34(1):39–48.

Chapter 8
The Social Contract and Alzheimer's

The Government's Priorities in Protecting its Citizens from Harm and Disease

Don't worry too much about terrorism. According to the National Safety Council, your odds of dying from a terrorist attack in the USA are 1 in 20 million. While the US government raises its defense budget by billions of dollars every year to protect you from this terrorist threat, or possibly a nuclear attack from the evil empire, consider the greater danger of dying from being struck by lightning (odds 1 in 5 million) or drowning in a bathtub (odds 1 in 800,000). Somehow, the US government in its mystifying reasoning has decided that infinitely more money should be budgeted to keep terrorists from harming you than to put that money to find a cure or prevention for any disease known to man that can and will more readily kill you.

For example, in 2014, if you were fortunate enough to avoid or recover from a heart attack, cancer, stroke, or other lethal malady to reach age 65, your odds of dying from Alzheimer's disease was 1 in 9 [1]. Despite this intimidating statistic, there is good news and bad news.

The good news is that every year after age 65, your chances of dying from any illness except Alzheimer's has improved substantially compared to 25 years ago. This is largely due to significant advancements in medicine and technology, as well as more effective drug therapy to control or prevent life-threatening conditions. Better health care provided by medical discoveries and technological innovation has given life expectancy a significant increase for the past 50 years in the USA. The medical advancements that have increased actuarial age in most modern countries from age 52 in 1960 to age 70 in 2010 could not have been achieved without the modest monetary investment by government and private industry to medical research. It is anybody's guess how much more health care may have improved for the same period of time if only half the defense budget had been allocated to medical research in the USA.

J.C. de la Torre, *Alzheimer's Turning Point*,
DOI 10.1007/978-3-319-34057-9_8

Now for the bad news. The chance of developing Alzheimer's for an individual who reaches age 65, doubles every five years, until by age 85, the odds of getting this dementia reach an alarming 50 % probability [1].

Dollars Per Death Index

Let us take a look at how government officials from the US Department of Health & Human Services view Alzheimer's research. They put out the following message. "As the leading funder of Alzheimer's research, the federal government is supporting significant new research into the causes of Alzheimer's and finding ways to delay, prevent, or treat the disease."

When the facts and figures are examined, the political and medical reality appears divergent. If you seriously care about ways to dodge getting Alzheimer's, where would you prefer to see your tax dollars go to, National Defense to protect you from a terrorist attack or Alzheimer's research to prevent or reduce your risk of dying from dementia?

The "dollars per death" index may be helpful in answering this question. Dollars per death refers to how many dollars the US government spends on each patient with a particular deadly disease. In the case of Alzheimer's (the sixth major cause of death in the USA), the National Institutes of Health doled out $586 million for research projects funded in 2015. This amounts to about $117 per patient for the 5 million individuals living in the USA with this dementia. It is anticipated Alzheimer's will strike 14 million people by 2050 [1]. The cost of care for a single Alzheimer patient is now estimated to be about $77,000 per year. This price tag includes prescription drugs, adult day care or home care services, personal care supplies, full time care at a facility. In 2015, the societal cost of Alzheimer's is estimated to be $226 billion. Since Alzheimer's is a progressive disease that may last 10 or more years and its incidence is rapidly rising every year, this annual cost per patient quickly adds up to astronomical figures. The societal cost of Alzheimer's is expected to top $1 trillion by 2050 and many economists warn this figure will devastate Medicare and Medicaid (Fig. 8.1). Despite a budget increase to NIH for fiscal year 2016 of $2 billion approved by Congress, medical research remains on the brink of collapse. The reason is that the cost of living expenses have risen considerably in the USA during the past decade and this rise has not kept pace with any significant increases in the NIH budget since 2003. A 12 billion dollar increase in the NIH budget would have avoided the precarious state of biomedical research insecurity.

Consider now how much the USA spends on protecting its citizens from foreign attacks. Since the Soviet Union collapsed in 1999, the major enemy of the USA is now a band of medieval barbarians with automatic rifles and dilapidated Toyota pickup trucks. For this purpose, the US Military has bases in 63 countries where it employs 325,000 military personnel deployed in 800 military bases around the world. These numbers do not include 6,000 military bases and 1.5 million military

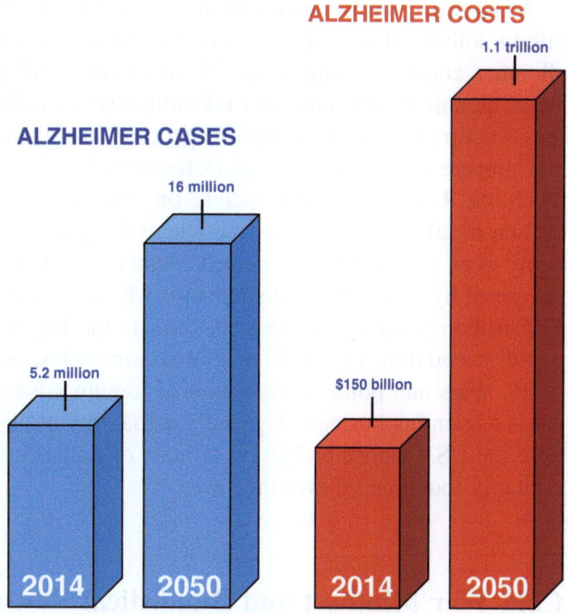

Fig. 8.1 Projected rise in the number of new Alzheimer cases and costs from 2014 to 2050

personnel located in the USA. These foreign military bases include 56 facilities in Germany, 113 in Italy and 84 in Japan. The logic of this foreign invasion by US military in friendly countries has never been fully explained. Could this be some form of indirect foreign aid to the three countries with the largest world economies?

Imagine if you can, the military toys bought by the Defense Department in recent years. The concept of producing state-of-the-art jet fighters to use against terrorist groups that could menace the USA would be laughable if it were not tragically stupid. For example, the Defense Department in 2009 purchased 188 jet fighters F-22 from manufacturer Lockheed Martin Corporation and paid 412 million for each for a total cost of $77 billion. The Pentagon said the F-22 was too advanced for low-tech enemies such as ISIS or Al-Qaida, so it grounded them, perhaps waiting for a knife fight with Russia someday. Had the Pentagon officials and Congressional politicians really cared for protecting the public from death, they could have tripled the funding for Alzheimer's research by buying 186 F-22 jet fighters, two less than originally planned.

What is ironic is that over the years, F-22 pilots have reported dozens of incidents with the F-22 s in which the oxygen system malfunctioned, causing lightheadedness and confusion. One F-22 pilot died in 2010 when his oxygen system failed.

Another spirited program was the Pentagon's contract to build Northrop Grumman B-2 Spirit, a strategic bomber with a cost label of $2.1 billion. The Defense Department ordered 21 of these and ended the contract to build more because, according to a spokesperson, "they were too expensive."

As of 2015, the National Defense budget in the USA is estimated to be 786.6 billion dollars. This budget pays in part for 20 modern aircraft carriers (more than all other countries combined), 72 submarines, 62 destroyers, 2,700 fixed-wing attack aircraft, 8,800 tanks, and 1.4 million active military personnel. This arsenal is only the tip of the iceberg with respect to our military expenses.

Compare the annual National Defense budget figure (about $1 trillion) with how much the US government spends on Alzheimer research or prevention ($585 million in 2015) and one can see where the priorities lie. This insane attitude has rarely been challenged by elected members of Congress for fear of not getting re-elected by a vengeful arms industry whose business with the United States tops $1.5 trillion yearly. The USA, in turn, is the largest exporter of weapons in the world, accounting for 30 % of global arms sales, according to a recent Swedish study. Wars and political instability of foreign governments make US arm exportation a booming business. Cynically, a majority of members in Congress cheerfully raise the US Defense budget by billions of dollars every year to demonstrate how militarily tough and clever they are.

Cold War Mind-set and Biomedical Research

Politicians in Congress consider it their "solemn duty" to protect you against terrorism in US soil and give that as the main reason for their eagerness to increase the Defense budget every fiscal year. But when it comes to Alzheimer's, that concern for your welfare and health fades into the night. Why is that? The reason is that these **congressional district and state representatives have one major concern: getting re-elected.** A good way to do this is by keeping the voters happy in their state or municipalities by bringing in money for jobs in the weapons industry, national security, military bases, and other jobs related to national defense. A second way to get re-elected is to have a chestful of campaign money, usually filled by lobbyists from defense contractors. The bottom line is that politicians elected to Congress do not think about the welfare of the people because it is unlikely that all the F-22 s, submarines, flattops, or destroyers will save more Americans than will die from Alzheimer's in the next 20 years. Since it is ultimately the people who elect these officials to go to Congress, it seems an irony that at the time of re-election, no accounting or rejection of these political charlatans is considered by the voters who must live with diseases that can be mastered. A number of representatives elected to the Senate and House maintain a Cold War mind-set which no amount of logic can dispel.

If these elected officials spent 3 months caring for an Alzheimer patient 24/7 instead of engaging in pointless political diatribes, they would reduce their endless adversarial posturing and make it their goal to prevent biomedical research from crumbling for lack of funding. This obvious lack of understanding by Congress to protect people from disease has forced NIH to cut funding for 85 % of all biomedical research projects submitted. When biomedical research projects are denied funding, graduate students, postdoctoral fellows and established researchers are forced to find other work outside of academic research. How in the name of human decency can Congress allow this to continue?

On the other hand, the annual costs by the Defense Department are only a slender piece of the pie when the recent wars in Iraq and Afghanistan are considered. The Congressional Research Service, a branch of Congress that provides bipartisan legal analysis and policy issues to members of Congress, estimates that these wars have cost the American taxpayer $1.6 trillion. However, a fuller accounting of the two wars which includes long-term care and disability compensation for veterans and families, military replenishment and socioeconomic costs puts the total cost at $4–6 trillion.

Does the monumental military spending really keep us safe from terrorism? If 9/11 is any indication, it is safe to assume that all the aircraft carriers, attack aircraft, submarines, tanks, and military intelligence did nothing to prevent 19 terrorists from killing nearly 3,000 Americans. There seems to be an unexplainable dilemma in how politicians decide the priorities that will keep us safe and alive.

It is important to point out that in 2014, Alzheimer's cost to Medicare and Medicaid was $150 billion, a figure that will balloon to $1.1 trillion by 2050 as longer life expectancy and aging baby boomers become major candidates to develop Alzheimer's (Fig. 8.1).

Clearly, something needs to be done now, not later, if we are to deconstruct, decelerate, and manage our worst enemy within: the consummate mind annihilator.

Reference

1. Alzheimer's Association. Alzheimer's disease facts and figures. Alzheimer's and Dement. 2014;10 (2): 16–20.

Chapter 9
Genetics of Alzheimer's Disease

Gene Mutations as the Cause of Alzheimer's

To understand how late-onset, also known as sporadic Alzheimer's disease, differs from early-onset or familial Alzheimer's disease, it is important to discuss the genetics of the latter. Genetic research has shown a link between genes on chromosomes 1, 14, and 21 and familial Alzheimer's. A gene on chromosome 19 called ApoE4 is associated with sporadic Alzheimer's and is considered a "susceptibility" gene because even when it is inherited from one or both parents, it can only increase the risk for Alzheimer's, not its certainty. Although sporadic and familial Alzheimer's have a similar clinical course, the familial type tends to develop severe brain atrophy and cognitive impairment around age 45; patients can develop muscle twitches and spams called myoclonus during the disease. The most important difference between sporadic and familial Alzheimer's is the specific gene mutation seen in the latter which directly assures that Alzheimer's will develop. In addition, familial Alzheimer's will affect individuals at a much earlier age, between 40 and 60 as opposed to over age 65 in most sporadic Alzheimer's cases. Diagnosis of familial Alzheimer's can be done by noting family history and neurological symptoms and observing severe brain atrophy on neuroimaging, by age and by genotyping.

A brief review of how genes are affected in familial Alzheimer's can provide a background for better understanding this dementia. Genes are located in chromosomes that are tightly packed within deoxyribonucleic acid (DNA) inside the nucleus of every cell. Genes are responsible for providing DNA with all the instructions it needs to carry on with its job of making proteins that will sustain life. As the keeper of genetic information in each living cell, the integrity and stability of DNA are essential to life. DNA is composed of building blocks called nucleotides that consist of a deoxyribose sugar bound on one side and a phosphate group bound to the other side of four nitrogenous bases. The bases are composed of four amino acids: adenine (A), thymine (T), guanine (G), and cytosine (C). A and T always pair together as does C and G (Fig. 9.1). Damage to DNA and its code of instructions

J.C. de la Torre, *Alzheimer's Turning Point*,
DOI 10.1007/978-3-319-34057-9_9

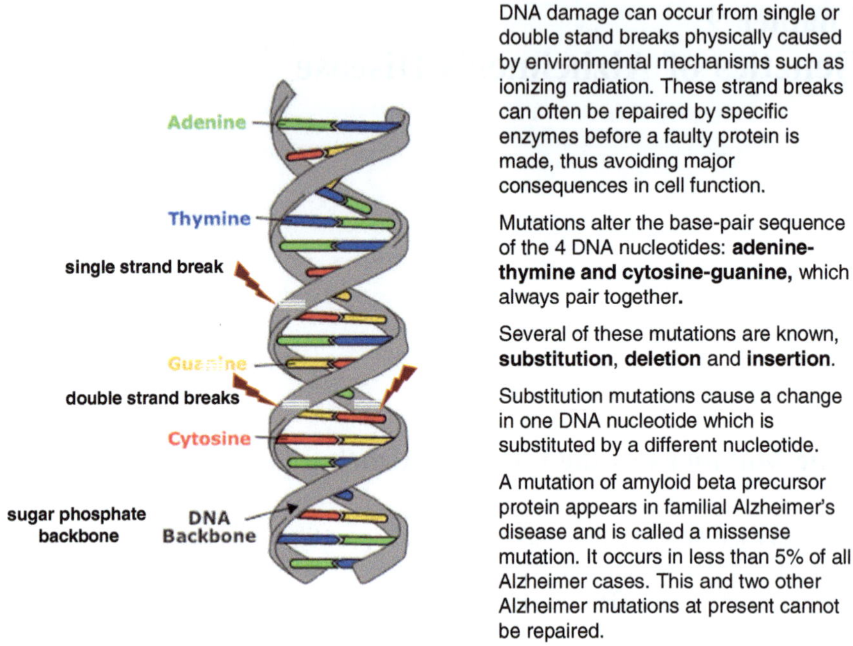

Fig. 9.1 DNA damage and mutation

for making proteins can occur in a number of ways that affect or change the normal rule of A-T and C-G sequences.

Like any other biologic molecule, **DNA is continually the target of damage from the environment** which, if not repaired, can result in metabolic dysfunction and cell death (Fig. 9.1). By contrast, **DNA mutations are inherited** from one or both parents or influenced by environmental factors including tobacco, ultraviolet light, ionizing radiation, X-rays, plant toxins, viruses, and many industrial chemicals. The end result of these mutations is faulty genes and damaging or lethal protein production. With faulty gene inheritance, any slight change in the DNA code of instructions can produce an abnormal protein that can lead to cell malfunction, disease or death. These malfunctioning proteins can be inherited from one or both parents. This is what happens in chromosomes 14, 1, and 21 when a single-gene mutation from one or both parents to their child leads to familial Alzheimer's disease later in life.

Familial Alzheimer's Disease

Because familial Alzheimer's is an autosomal dominant disorder, inheriting a mutated gene from one parent results in a 50 % chance that the dementia will develop early, that is, from age 40 to 60, as opposed to sporadic Alzheimer's which usually begins after age 65. If unlucky, the child that inherits the mutated gene from both parents has a 100 % chance of developing familial Alzheimer's. This is a very rare occurrence. Familial Alzheimer's affects less than 5 % of all Alzheimer's cases, and it is estimated to affect less than 200,000 people in the USA compared to 5.3 million cases of sporadic Alzheimer's as of 2015 [1].

The gene mutation for familial Alzheimer's produces PS-1 and PS-2 in chromosomes 14 and 1, respectively, and APP in chromosome 21. All three mutated genes lead to an increased amount in the production of Abeta peptide which is an abnormal snip-off fragment of APP. In time, the Abeta contributes to the buildup of tissue clumps called senile plaques which appear characteristically in the Alzheimer's brain.

It must be emphasized that only a small percentage of gene mutations cause genetic disorders and most have no impact on the health or mental state of an individual. Thousands of DNA toxins from the environment that could cause a disease are often repaired daily by a variety of repair strategies used by the DNA molecule to prevent a faulty protein from being produced or expressed.

That is also true for some mutations that can change a gene's DNA sequence but do not change the function of the protein that is produced by the gene. Mutations causing disease can occur both from inheritance and from environmental sources. Tobacco, ultraviolet light, and other toxic chemicals are all potential enemies of DNA. One way these hazards can attack our genes is very sneaky. They have the ability to break up the chemicals making up the DNA molecule. For example, some mutagens like to **substitute**, **delete,** or **insert** an extra chemical to the DNA molecule. When these mutations occur, DNA replicates the gained abnormalities into millions of cells causing major damage or death.

It is not known what triggers the mutation of genes responsible for the Abeta production in familial Alzheimer's **although risk factors for sporadic Alzheimer's disease may provide some clues**. For example, cerebral amyloid angiopathy is a condition where Abeta accumulates inside the brain vessels where it can impair blood flow delivery to the brain. The cause of APP mutation and Abeta elaboration has been linked in some cases of familial Alzheimer's to strokes and amyloid angiopathy [2]. Stroke is a major risk for sporadic Alzheimer's and in rare circumstances to familial Alzheimer's.

Two studies have looked at the role of cerebral blood flow in familial Alzheimer's disease. The first study from Harvard neurologist **Dr. Keith Johnson** and his colleagues [3] looked at familial Alzheimer's patients with PS-1 mutation from a family in Colombia suffering from this dementia. They measured and compared brain blood flow in the familial Alzheimer's patients with family members who were carriers of PS-1 but who did not yet show symptoms of the disease and with cognitively normal non-PS-1 carriers. These investigators found that cerebral blood flow was reduced in the asymptomatic PS-1 carriers in various regions of the brain including the hippocampus, cingulate cortex, and posterior parietal lobe, regions that are known to accumulate neurodegenerative matter associated with Alzheimer's. The PS-1 patients expressing Alzheimer's symptoms also showed similar reduced blood flow to the brain as the asymptomatic individuals, while the cognitively intact individuals showed normal cerebral blood flow levels [3].

In the second study by **Sebastien Verclytte** and his group in Lille, France, it was shown that people with familial Alzheimer's disease develop reduced blood flow in specific areas of their brain that are linked to cognitive decline and dementia. These low perfused brain regions also show an impairment in neuronal metabolism indicated by reduced penetration of glucose into nerve cells, a condition called *reduced glucose uptake* and a key step in the production of brain cell energy [4]. The brain regions where severe low perfusion was observed in a small population diagnosed with familial Alzheimer's disease are the cingulate cortex, frontal lobes, and parietal lobes which are also the areas that suffer the most severe damage in non-genetic Alzheimer's [4].

What is striking about these studies is that it provides an important clue that may help explain how blood flow insufficiency in key areas of the brain may trigger the PS-1 gene mutation responsible for familial Alzheimer's and possibly for the other familial Alzheimer's mutations. Since in the first study the asymptomatic PS-1 carriers did not show cognitive impairment at the time of cerebral blood flow measurement, it can be assumed that neurodegeneration was relatively minor or absent from their brain, and otherwise, measurable memory loss and cognitive deterioration would have been observed. In the second study where cerebral blood flow and metabolism were measured, the reduced brain perfusion may have been an extension of the brain hypoperfusion seen in the asymptomatic patients carrying the PS-1 gene mutant. Although this observation is purely hypothetical, it may provide a clue for further research into the question of "at what point do patients carrying the gene mutation for familial Alzheimer's begin to lower their cerebral blood flow and metabolic demand for energy nutrients?" Followed by this whimsical question, "does cerebral blood flow trigger familial Alzheimer's or is it just an innocent bystander?"

Unfortunately, in the Johnson study, no follow-up was carried out to see how many of the asymptomatic PS-1 carriers eventually converted to familial Alzheimer's, an opportunity which might have provided crucial information on the role of deficient brain blood flow in the expression of this inherited dementia.

Twin Studies

The study of disease using identical and fraternal twins has been useful since the nineteenth century in trying to analyze the role of the environment with heredity. Identical twins are formed from a single ovum fertilized from a single sperm which gives rise to two separate fetuses that share the same gender and all their genes. Fraternal twins are formed from two different ova fertilized by two different sperm and essentially the fetuses are different siblings born at the same time. The idea that fraternal twin studies provide information regarding the effect of environment on their behavior is based on the assumption that both twins grew up in the same place and were exposed to the same things until a certain age. Identical twins in addition to the environmental similarities shared by fraternal twins may show dissimilarities, but these must be caused by environment not genes. There is a reasonable logic to this assumption but also considerable room for error when some conclusion is made.

Although not immune to criticism from workers in the field, the classical twin study design relies on comparing monozygotic (identical) twins with dizygotic (fraternal) twins to see the full effect of genetic mechanisms to the direct effects of the environment on the expression of a disease. This **gene–environment interaction** has been used repeatedly to study Alzheimer's disease.

Since it is already known that there is a definite role for at least three genes (PS-1, PS-2, and APP) that can mutate and promote familial Alzheimer's, twin studies have focused on the possible genetic links that may exist in sporadic Alzheimer's.

One such study compared identical twins raised in different environments on the possible development of sporadic Alzheimer's later in life [5]. A marker (5-hydroxymethylcytidine or 5-hmC) that is metabolically altered in age-related neurodegenerative disorders such as Alzheimer's was observed to be significantly diminished in the brain of one of the two identical twins who had died of Alzheimer's but not in the other twin who remained cognitively healthy [5]. The marker that was found diminished in the Alzheimer's twins is likely the product of some environmental stimulus on the genetic makeup of some individuals because it did not affect DNA sequence in the manner that produces missense mutation in familial Alzheimer's. Instead, **the marker reflects DNA** damage **during** life in those persons who are likely exposed to environmental conditions that can lead to Alzheimer's dementia.

The term **"discordant twins for Alzheimer's"** is commonly used to refer to identical twins where one of the twins develops Alzheimer's in his or her lifetime but not the other [6]. Most of the evidence on the subject indicates that a risk to Alzheimer's disease in monozygotic and dizygotic twins suggests that the association is a non-heritable genetic aberration acquired during aging or in the presence of Alzheimer's risk factors [7, 8]. This is especially true in identical twins who share the same DNA.

This view is not shared by proponents of the Abeta hypothesis who posit that Abeta formation in the brain is caused by APP mutation. This belief has led to the widely held tenet that sporadic Alzheimer's can occur from a mutation of APP giving rise to Abeta accumulation in the brain which ostensibly causes neurodegeneration, cognitive loss, and other Alzheimer's pathology. However, there is no evidence to suggest that single-gene mutations directly responsible for familial Alzheimer's disease are also involved in sporadic Alzheimer's and a number of identical twin studies have supported this conclusion.

Among identical twins who share the same genetic makeup, Alzheimer's disease in one twin does not mean that the other twin will inevitably get the disease. Studies have shown that among male identical twins, when one twin develops Alzheimer's disease, the other twin develops the disease only 45 % of the time. This finding assumes that if Alzheimer's were strictly a genetic disorder, as it is in familial Alzheimer's, both twins should contract the disease about the same time as the other. However, when both identical twins did develop Alzheimer's, there was as much as a 15-year difference in age of onset, suggesting that environmental factors played a major role in the development of Alzheimer's [9].

It seems reasonable to suspect that a gene–environment interaction contributes to both early- and late-onset Alzheimer's disease. An example of this gene–environment interaction is seen in certain types of cardiovascular disease that can pose a major risk factor for Alzheimer's [10]. One cardiovascular risk of Alzheimer's which can affect blood flow to the brain during aging is the size and shape of the heart's left ventricle. It was reported that in identical twins, left ventricular hypertrophy (LVH), a condition where the size of the ventricle is bigger than normal, was similar in a large percentage for both twins. The study also found that in a smaller percentage of the same identical twin pairs, LVH differed significantly [11]. This finding suggests that both genetic and, to a lesser extent, environmental factors showed a marked effect on ventricular size and shape in identical twins. Thus, genetics and environment could independently increase or lower the risk of Alzheimer's in identical twins, at least with respect to this cardiac anomaly.

This gene–environment interaction is also likely to affect other medical conditions which can directly influence LVH, such as high blood pressure, heart valve disease, aortic stenosis, obesity, and diabetes, conditions which are curiously potent risk factors for Alzheimer's disease [12, 13].

The gene–environment interaction has led to the fast growing field of epigenetics, an area of research that is presently being applied to the search of early diagnostic tools for Alzheimer's and possibly novel treatments to prevent or better manage this dementia. Epigenetics involves processes that are known to be dynamic and reversible, unlike static and irreversible genetic disorders such as familial Alzheimer's.

Susceptibility Genes, Can They Harm You?

A number of **susceptibility genes** have been described that appear to increase the risk of contracting sporadic Alzheimer's disease. Susceptibility genes differ from the early-onset mutant genes (PS-1, PS-2, and APP) in that their presence does not necessarily mean that even if inherited from both parents, they will definitely cause Alzheimer's dementia.

So far, 11 susceptibility genes to Alzheimer's dementia have been identified, but their exact function in the pathologic cascade that characterizes loss of memory has not been clearly revealed. **It is not important to know whether you carry any of these susceptibility genes for two reasons. First, even if you carry one of the nasty ones, you may not get Alzheimer's, and second, if you do get Alzheimer's, there is no treatment to reverse or prevent it.** Knowing that you are carrying a susceptibility gene consequently can only be a source of frustrating and pointless anxiety.

Fortunately, many of the susceptibility genes are extremely rare in the general population. Some of these genes are suspected of playing a role in energy metabolism, protein degradation, inflammation, and the immune system [14]. We anticipate that some susceptibility genes will be linked to small blood vessel abnormalities in the brain, heart, kidney, and eye retina, in fact, anywhere where blood flows. Much information has been gathered and continues to be reported that these organs are quite vulnerable to poor blood flow during advanced aging [15–18].

The most studied of the susceptibility genes is the apolipoprotein E (ApoE), derived from the ApoE gene encoded on chromosome 19. The ApoE gene has a role in lipoprotein metabolism in brain and is known to influence cardiovascular, neurological, and infectious diseases [19]. Three different forms of this gene have been described: ApoE2, ApoE3, and ApoE4.

The only susceptibility gene variant associated with sporadic Alzheimer's is the gene encoding ApoE4. ApoE4 is the subject of intense investigation due to its negative effect on the development of Alzheimer's. If ApoE4 **is inherited from one parent, it doubles the risk of contracting Alzheimer's, and the risk is much greater in women than in men** [20]. The number of women with Alzheimer's far exceeds that of men but not necessarily because of the greater prevalence for the ApoE4 gene. Women tend to live longer than men and consequently have a greater opportunity of developing dementia. There may be other factors involved in women's higher risk of developing Alzheimer's dementia when compared to men, such as estrogen levels, but the evidence for this occurrence is only suspect at this time. The reason for this gender vulnerability is unclear but is the subject of heightened investigation.

A person normally inherits two alleles or copies for each trait, one from each parent. **If a person inherits an ApoE4 allele from both parents carrying the gene, the risk of Alzheimer's rises 10-fold**. However, not everyone who inherits ApoE4, even from both parents, will get Alzheimer's although 60–80 % of all Alzheimer's patients appear to carry one or both copies of this gene.

ApoE2, by contrast, appears to have a protective effect against acquiring Alzheimer's, while carrying ApoE3 has no effect either way. The presence of only one ApoE4 allele is found in about 15 % of the population, while those carrying one copy of the ApoE2 are about 5 %. Some people who carry **a copy of ApoE4 and a copy of ApoE2 will have a lower risk of developing Alzheimer's**, so it would appear that the negative effects of ApoE4 are more overpowered by the presence of ApoE2.

The increased risk period for Alzheimer's is at its greatest between ages 60–80, suggesting that acquiring Alzheimer's at a very late age has a relatively small genetic component to it and perhaps influenced by the type of ApoE alleles being carried.

Although it is not clear how ApoE2 confers some degree of protection against Alzheimer's disease, recent evidence suggests that inheriting one or both ApoE2 alleles appears to protect against cardiovascular disease, a major risk factor for Alzheimer's. This allele also seems to confer some protection against early atrophy of the hippocampus, a brain region that is violently attacked by neurodegeneration prior to cognitive meltdown. Early hippocampal atrophy is a common preclinical finding in people who later develop cognitive decline and Alzheimer's. Another reason for ApoE2 in providing relative protection against Alzheimer's dementia may be its positive influence on mitochondrial dynamics, the organelle responsible for generating the energy needed by neurons to stay alive. If this finding is confirmed, the next research step is to determine whether inheriting an ApoE2 allele in contrast to an ApoE3 allele slows down the rapid cognitive decline that is typified in Alzheimer's dementia.

What seems evident from the research at this time is that better understanding of the Apo genotype and their influence on cognitive function is fundamental to putting the pieces of the Alzheimer's jigsaw puzzle together to see how this dementia can be controlled.

References

1. Alzheimer's Association. Alzheimer's disease facts and figures. Alzheimer's Dement. 2014;10 (2):16–20.
2. Rossi G, Giaccone G, Maletta R, Morbin M, Capobianco R, Mangieri M, Giovagnoli AR, Bizzi A, Tomaino C, Perri M, Di Natale M, Tagliavini F, Bugiani O, Bruni AC. A family with Alzheimer disease and strokes associated with A713T mutation of the APP gene. Neurology. 2004;63(5):910–2.
3. Johnson KA, Lopera F, Jones K, Becker A, Sperling R, Hilson J, Londono J, Siegert I, Arcos M, Moreno S, Madrigal L, Ossa J, Pineda N, Ardila A, Roselli M, Albert MS, Kosik KS, Rios A. Presenilin-1-associated abnormalities in regional cerebral perfusion. Neurology. 2001;56(11):1545–51.
4. Verclytte S, Lopes R, Lenfant P, Rollin A, Semah F, Leclerc X, Pasquier F, Delmaire C. Cerebral hypoperfusion and hypometabolism detected by arterial spin labeling MRI and FDG-PET in early-onset Alzheimer's disease. J Neuroimaging. 2016;26(2):207–212.
5. Chouliaras L, Mastroeni D, Delvaux E, Grover A, Kenis G, Hof PR, Steinbusch HW, Coleman PD, Rutten BP, van den Hove DL. Consistent decrease in global DNA methylation and hydroxymethylation in the hippocampus of Alzheimer's disease patients. Neurobiol Aging. 2013;34(9):2091–9.
6. Renvoize EB, Mindham RH, Stewart M, McDonald R, Wallace RD. Identical twins discordant for presenile dementia of the Alzheimer type. Br J Psychiatry. 1986;149:509–12.
7. Forsberg LA, Absher D, Dumanski JP. Non-heritable genetics of human disease: spotlight on post-zygotic genetic variation acquired during lifetime. J Med Genet. 2013;50(1):1–10.
8. Räihä I, Kaprio J, Koskenvuo M, Rajala T, Sourander L. Environmental differences in twin pairs discordant for Alzheimer's disease. J Neurol Neurosurg Psychiatry. 1998;65(5):785–7.
9. Räihä I, Kaprio J, Koskenvuo M, Rajala T, Sourander L. Alzheimer's disease in Finnish twins. Lancet. 1996;347:573–8.
10. de la Torre JC. Cardiovascular risk factors promote brain hypoperfusion leading to cognitive decline and dementia. Cardiovasc Psychiatry Neurol. 2012;Article ID 367516, 2012:15p.
11. Harshfield GA, Grim C, Hwang C, Savage D, Anderson S. Genetic and environmental influences on echocardiographically determined left ventricular mass in black twins. Am J Hypertens. 1990;3:538–43.
12. de la Torre JC. Is Alzheimer's disease a neurodegenerative or a vascular disorder? Data, dogma, and dialectics. Lancet Neurol. 2004;3(3):184–90.
13. de la Torre JC. Cerebral hemodynamics and vascular risk factors: setting the stage for Alzheimer's disease. J Alzheimers Dis. 2012;32(3):553–67.

14. Escott-Price V, Bellenguez C, Wang LS. Choi SH Gene-wide analysis detects two new susceptibility genes for Alzheimer's disease. PLoS ONE. 2014;9(6):e94661.
15. O'Rourke MF, Safar ME. Relationship between aortic stiffening and microvascular disease in brain and kidney: cause and logic of therapy. Hypertension. 2005;46(1):200–4.
16. de la Torre JC. How do heart disease and stroke become risk factors for Alzheimer's disease? Neurol Res. 2006;28(6):637–44.
17. Cheung CY, Ong YT, Ikram MK, Chen C, Wong TY. Retinal microvasculature in Alzheimer's disease. Alzheimers Dis. 2014;42(Suppl 4):S339–52.
18. Jefferson AL, Beiser AS, Himali JJ, Seshadri S, O'Donnell CJ, Manning WJ, Wolf PA, Au R, Benjamin EJ. Low cardiac index is associated with incident dementia and Alzheimer disease: the framingham heart study. Circulation. 2015;131(15):1333–9.
19. Mahley RW, Weisgraber KH, Huang Y. Apolipoprotein E: structure determines function, from atherosclerosis to Alzheimer's disease to AIDS. J Lipid Res. 2009;50(Suppl):S183–8.
20. Altmann A, Tian L, Henderson VW, Greicius MD. Alzheimer's disease neuroimaging initiative investigators. Sex modifies the APOE-related risk of developing Alzheimer disease. Ann Neurol. 2014;75(4):563–73.

Chapter 10
Powering the Brain

The Brain Needs Constant Energy Nutrients to Stay Healthy

The mammalian brain is a remarkable organ, but the human brain is even more unique in many ways. All animals can move, feel, see, hear, smell, remember, and make decisions, but only humans can use language and writing to communicate, as well as display a capacity for abstract thinking, scientific inventions, wisdom, logic, creativity, and the creation of a complex social network.

Although the human brain is only 2 % of the total body weight, it requires 20 % of the body's oxygen and 25 % of the body's glucose to create the energy needed for brain cells to stay alive. Mitochondria are the source of energy for all cells and are especially important n neurons where one of the primary functions is to match the supply of energy need to the demand. In neurons, mitochondria can be found in the cell cytoplasm and their primary role is to manufacture adenosine triphosphate (ATP) by systematically metabolizing energy from nutrient molecules derived from food. ATP is the currency used by cells to yield energy for every task needed by enzymes, proteins, and cellular functions. Life is not possible without ATP.

Mitochondria have been called the power houses of cells but they are more like batteries, where energy nutrients are stored and dispensed as needed by the cell. The main nutrients required to produce cell energy are glucose and oxygen. These two molecules are delivered to neuronal mitochondria via the cerebral circulation. Everything in the body operates on energy. The formation of new cells, waste disposal, muscle movement, wound healing, smiling and thinking, all require energy.

The brain requires a continuous and uninterrupted supply of glucose and oxygen because it cannot manufacture either substance. In normal circumstances, glucose is the sole energy fuel for the brain. For brain cells to receive their needed energy supply, both glucose and oxygen are delivered to the brain by capillaries that are located near the brain cells. Once transported inside the cell, the energy production

J.C. de la Torre, *Alzheimer's Turning Point*,
DOI 10.1007/978-3-319-34057-9_10

continues as glucose undergoes several major changes. These changes yield to the formation of *glycolysis*, the *citric acid cycle* (also called tricarboxylic acid cycle or Krebs cycle), and the *electron transport chain.*

These 3 major reaction pathways lead to the creation of ATP. The process to produce such energy is called cellular respiration and involves a series of metabolic reactions inside the cell to produce ATP energy. Why is so much energy needed by the brain tissue? Many studies indicate that two-thirds of the brain's energy budget is used to help neurons transmit signals to other neurons in the brain, a process called neurotransmission, or to accomplish tasks in other tissues of the body, for example, in sensation or movement. The other third of the energy supply is used for maintaining the cell's health and survival [1].

When glucose and oxygen supply to the brain is progressively diminished due to reduced brain blood flow, (which commonly happens in unhealthy aging), brain cell activities can decline or not function at all. Poor blood flow to the brain and capillary distortions also reduce waste products from exiting the blood–brain barrier. This may cause Abeta, normally formed in the brain, to accumulate and become toxic within the brain tissue by preventing its exit as a waste product. The blood–brain barrier is a permeable barrier in capillaries that selects which substances will enter and exit the brain. Other waste products besides Abeta that cannot exit the brain are bound to accumulate and may cause toxic reactions in the brain tissue.

When mitochondrial function to carry out the complex energy-demanding processes involved in neurotransmission is negatively affected, many nerve cells cease to communicate with each other (Fig. 7.1). The effect is reflected in frequent 'senior moments' that, depending on the number and location of the nerve cells affected, may or may not lead to cognitive impairment and dementia.

The slow and chronic dwindling of blood flow to the brain during the aging process is known as **cerebral hypoperfusion**. It is an important phenomenon experienced by virtually everyone who ages and people of any age that develop brain disease or brain trauma. **It does not necessarily lead to Alzheimer's or other dementias unless brain blood flow diminishes to a critical level where neurons can no longer sustain their survival**. When that happens, a **neuronal energy crisis** can follow which triggers the carving up of ischemic-sensitive neurons, in the hippocampus and posterior parietal cortex, regions where memory and cognitive functions reside [2] (Fig. 7.1).

Neuronal Energy Crisis

The metabolic meltdown resulting from a **neuronal energy crisis** sets up the beginning of cognitive deterioration, progressive neurodegeneration, and atrophic changes in the brain (Fig. 7.1).

Prevention of that critical level of brain blood flow compromise depends on its detection and intervention. This topic is reviewed in Chap. 15.

Brain mitochondria can also be compromised by a sudden event such as a stroke or heart attack. The cognitive failure after an ischemic stroke or heart attack can quickly appear within minutes or hours and may be reversed or prevented with dramatic medical intervention such as with the intravenous use of tissue plasminogen activator (tPA) which must be given within 3 h after the symptoms of stroke appear and within 12 h after a heart attack.

The most serious effect of a neuronal energy crisis is the inability by the neurons affected to normally respond to neural activation brought on by stimuli such as thinking, reasoning, remembering, and task planning. Loss of the universal cell fuel ATP will set up a domino effect which will impair protein synthesis, protein folding, and protein assembly and cause neurotransmission deficits. If ATP loss can be reversed or restored (see Chap. 18), there is a very good chance that Alzheimer's and other dementias will be prevented before their devastating clout can be expressed.

Making ATP Energy Fuel

Glycolysis is the main pathway for producing the cell fuel ATP from glucose, and it is the first stage of cellular respiration. Cellular respiration has nothing to do with breathing but is actually a process that generates energy for cells from foods we eat. It does this in the cell cytoplasm by breaking down glucose into 2 molecules of pyruvate which produces 2 molecules of ATP and 2 molecules of NADH, the reduced form of nicotinamide adenine dinucleotide (NAD^+) whose function in part is to transfer electrons in the electron transport chain process to make ATP [3]. Pyruvate is broken down into acetylcoenzyme A (acetyl-CoA) which initiates the citric acid cycle. Glycolysis at this stage of energy production does not require oxygen.

The Citric Acid Cycle (Tricarboxylic Acid Cycle or the Krebs Cycle)

In the second stage of cellular respiration, the citric acid cycle begins in the mitochondria, tiny organelles that perform like micropower plants to turn carbohydrates, proteins, and lipids derived from food to produce ATP. Through a series of chemical transformations, acetyl-CoA derived from pyruvate fuels this cycle. In the mitochondrial matrix, and with the help of oxygen, the waste products carbon dioxide (CO_2) and water are produced to generate usable energy. In addition, the cycle consumes acetyl-CoA and reduces NAD^+ to NADH. At this step, NADH (and another electron carrier, $FADH_2$) transfers electrons to oxygen to create oxidative phosphorylation, a fundamental process in the making of ATP. These reactions cannot occur if oxygen is absent during this cycle, for example from an anoxic (no oxygen) or severe hypoxic (low oxygen) event. When oxygen is unavailable or the citric acid cycle is blocked, there is a shift to the Embden Meyerhof pathway of glycolysis, where glucose is converted to lactic acid, a very inefficient way of producing ATP energy.

The Electron Transport Chain

The final stage of cellular respiration is characterized by the production of energy-rich ATP. The electron transport chain uses oxidative phosphorylation, a highly efficient aerobic energy generator, to release energy so that it can be converted to ATP. Normally, each molecule of citric acid at each turn of the citric acid cycle is able to generate 38 molecules of ATP to fuel brain cells. Lacking this energy supply is an extremely important misstep in the survival of neurons and glial cells. This is due to the fact that the lack of high energy supply required by the brain will create pathological consequences in the brain such as the intense and progressive formation of senile plaques and neurofibrillary tangles (or plaques and tangles), as well as memory loss, and other cognitive deficits.

Many things can happen that may impair glucose and oxygen from being delivered by the circulation to the brain. When the impaired delivery of glucose and oxygen occurs, specific parts of the brain responsible for recent memory, planning and performing daily tasks, logical thinking, use of language, and the ability to cope with daily social and working tasks will suffer and deteriorate. By the time Alzheimer's is diagnosed, even at the earliest stage, these functional activities will have worsened to the point where physiological and anatomical features of the normal brain will be markedly altered.

Further clues that the disease is dangerously nearing will be suggested by neuropsychological tests that measure memory function, daily planning skills, learning ability, and other features characteristic of psychological slowdown (see Chap. 15 **Psychometric testing**). In addition, some imaging tests to peer inside the brain can be used to compliment the psychological testing. These imaging tests use computerized tomography (CT scan), magnetic resonance imaging (MRI), and transcranial doppler (TCD). Findings from these tests may provide information about brain lesions or atrophic changes in the limbic system which are characteristics of dementia. These imaging tests can detect very subtle abnormalities involving metabolic brain changes and cerebral blood flow deficiencies that often precede Alzheimer's symptoms by many years [2, 4, 5].

Family members and friends may notice the person who is approaching an Alzheimer's diagnosis reach a stage between normal cognitive behavior and full-blown dementia. This stage is known as mild cognitive impairment and is characterized by personality changes such as increased anxiety, irritability, including verbal or physical outbursts, delusions, pathological jealousy, and depressive moods [6]. These early changes tend to target the affective domain, that is, feelings and emotions. The affective domain is able to express love, friendship, and affection and can display mood activities ranging from fear to aggression.

These emotions are mostly contained in the limbic system, a collection of brain structures which govern emotional behavior, memory, and motivation. The major center associated with the integration of emotions, emotional behavior, and motivation is the amygdala. The amygdala is an almond-shaped structure located in the temporal lobe of the brain. It is connected to many different regions of the brain including the prefrontal cortex, thalamus, septal area, hypothalamus, and hippocampus. These are areas where memory and emotions are combined. Besides integration of emotions, the amygdala plays a role in detecting fear and preparing for emergency events.

The amygdala can help store memories of events and emotions so that an individual may be alerted to recognize similar events in the future. For example, if a person falls hard while walking on an icy street, the amygdala will help recall that event and provide some cautionary thinking when walking on ice again. Stimulation of the left amygdala in humans causes intense emotions, such as aggression, sadness, anxiety, or happiness while stimulation of the right amygdala has a negative emotional role in the expression of fear and sadness [7].

Disturbances in mitochondrial energy production contribute to a variety of neurological and physical problems. Impaired oxidative and energy metabolism are often indicators of impending cognitive dysfunction that can convert to Alzheimer's disease.

I discuss in detail in Chap. 17 how critical factors that prevent efficient production of ATP can trigger the process of Alzheimer's dementia. This conclusion is based on substantial medical-based evidence that should cast doubt on old ideas about the cause of Alzheimer's and explains the litany of continued clinical trial failures seen in the last 25 years for this dementia.

How ATP energy is used in the brain may clarify the reason why poor brain blood flow targets neurons in cognitive-related regions but not so much in others.

Why Do Cognitive-Regulating Neurons Die During Aging?

We now know that Alzheimer's is not caused by a single pathological arrow striking some vital brain center, as a stroke or gunshot wound to the head can do. Instead, Alzheimer's disease targets specific regions of the brain with unbridled fury. The regional brain targets derived from Alzheimer's attack provides a clue as to where and how it will deliver its knockout punch. For example, **Dr. Daniel Silverman** of UCLA reported mapping regional brain territories in humans using FDG-PET scans and found that the rate of cerebral blood flow and glucose metabolism (a marker of energy consumption) are virtually identical [8]. One can then expect that when cerebral blood flow decreases in some brain region, metabolism also decreases in a parallel fashion to brain blood flow. This has been known for over a century and is referred to as coupling. More importantly, Silverman's FDG-PET study revealed that the cerebellum, despite its significantly lower metabolic activity compared to the hippocampus, is as richly perfused as that memory-storing limbic structure [8].

This finding is surprisingly compelling in explaining the physiopathologic changes that occur in the hippocampal neurons but not in the metabolically less active cerebellar neurons as Alzheimer's gets ready to strike. Because these two neuronal populations receive the same amount of blood flow but react metabolically differently to chronic brain hypoperfusion preceding Alzheimer's onset, it would appear that hippocampal neurons (the initial target of Alzheimer's pathology) need more energy supply than the more complacent or relatively less hardworking cerebellar neurons.

In my judgment, what may happen is that a neuronal crisis due to chronic poor blood flow to the brain selectively targets the ischemic-sensitive hippocampus (and other brain regions that regulate cognitive function) but inexplicably spares cerebellar neurons. **The neuronal crisis eventually induces the accumulation of Abeta-containing plaques and neurofibrillary tangles in neuronal populations that regulate cognitive function but strangely enough does not damage cerebellar neurons to the same extent. Why is that?**

This difference between abundant accumulation of plaques and tangles and neuronal supply and demand in the hippocampus or cerebellum infers that two pathologic events are occurring prior to Alzheimer's neurodegeneration. The first pathologic event sees Alzheimer's related vascular risk factors generating chronic brain hypoperfusion in all brain regions during aging. During this stage, **energy demand and consumption** is much greater in the hippocampal neurons than neurons in the cerebellum [8].

We do not know the exact reason why this happens but we suspect that the hippocampal pyramidal neurons are more ischemic-sensitive than cerebellar neurons because their metabolic energy dynamics at keeping memory and learning ability intact is greater than the motor tasks expended by cerebellar neurons.

Even though the glucose energy supply is similar to both hippocampal and cerebellar neurons, the cerebellar neurons remain relatively well-fed energetically, even with a deficient glucose delivery while the hippocampal neurons struggle to survive the glucose deficiency that is parsimoniously available to them from the brain hypoperfusion. This is not to say that the cerebellum is conspicuously spared of any Alzheimer's pathology but neurodegenerative damage is considerably less than in the hippocampus and senile plaques and neurofibrillary tangles which are abundant in the limbic system appear relatively rare in the cerebellum. Moreover, extensive neuronal damage to the hippocampus results in memory and learning deficits, the initial signs of Alzheimer's disease but cerebellar signs such as dysarthria, myoclonic jerks and ataxia are not seen until much later in the Alzheimer's pathologic process. As neuronal damage spreads to other limbic structures, the amygdala, a region that is involved in emotion, is affected, resulting in mood swings such as anger, irritability, and apathy. By contrast, movement-related functions such as walking or climbing stairs which are partially controlled by the cerebellum are not initially impaired until Alzheimer's worsens to the moderate-severe stages.

The brain cell energy crisis involving cognitive-related regions is the direct result of blood flow supply not meeting energy demand in highly active hippocampal neurons whose vascular reserve capability reaches an untenable critical threshold. **Stated another way, the energy-starved hippocampal neurons are unable to cope with the persistent brain hypoperfusion and consequently undergo oxidative and endoplasmic reticulum stress that markedly reduces the cell fuel ATP in a vicious cycle**. The molecular outcome that follows this process in the hippocampus negatively affects post-translation processing steps resulting in impaired protein synthesis, protein assembly, and protein folding. Because proteins do most of the work to keep a neuron healthy, defects occurring during their synthesis, assembly, or folding steps can compromise the normal intracellular and extracellular secretory transport pathway and threaten neuron survival in specific brain regions. Protein synthesis aberrations also lead to the excessive formation of Abeta-containing plaques. It should be pointed out that the concept of pyramidal neurons in the hippocampus working metabolically harder than cerebellar neurons is purely argumentative and not based on any solid evidence. Further research should provide better insights to this notion.

References

1. Swaminathan M. Why does the brain need so much power? Sci Am. 2008:29–30.
2. de la Torre JC. Pathophysiology of neuronal energy crisis in Alzheimer's disease. Neurodegener Dis. 2008;5(3–4):126–32.
3. Garrett HR, Grisham C. Biochemistry. Boston: Twayne Publishers; 2008.
4. Mosconi L, Mistur R, Switalski R, Tsui WH, Glodzik L, Li Y, Pirraglia E, De Santi S, Reisberg B, Wisniewski T, de Leon MJ. FDG-PET changes in brain glucose metabolism from normal cognition to pathologically verified Alzheimer's disease. Eur J Nucl Med Mol Imaging. 2009;36(5):811–22.
5. Xekardaki A, Rodriguez C, Montandon ML, Toma S, Tombeur E, Herrmann FR, Zekry D, Lovblad KO, Barkhof F, Giannakopoulos P, Haller S. Arterial spin labeling may contribute to the prediction of cognitive deterioration in healthy elderly individuals. Radiology. 2015;274 (2):490–9.
6. Weiss B. Alzheimer's surgery: an intimate portrait. Bloomington, Indiana: AuthorHouse; 2005.
7. Lanteaume L. Emotion induction after direct intracerebral stimulations of human amygdala. Cereb Cortex 2007;17(6):1307–13.
8. Silverman DH. Brain 18F-FDH PET in the diagnosisi of neurodegenerative dementias, comparison with perfusion SPECT and with clinical evaluations lacking nuclear imaging. J Nucl Med. 2004;4:594.

Chapter 11
Pharmaceuticals and Alzheimer's

Profit not Health Is the Pharmaceutical Mantra

Epidemiological studies throughout history have provided useful information on disease processes and sometimes breakthroughs in medicine that have easily saved millions of lives. For example, the cholera epidemic in nineteenth-century London was nearly wiped out from the single-handed study of **John Snow** who showed that the death rate from cholera was much higher in the zone where a public water pump was used by the community. After disinfecting the water and pumping in chlorine, the outbreak quickly ended.

More recently, a study **by Drs. Ernest Wynder** and **Evart Graham** in 1950 [1] revealed a cause–effect relationship between smoking and lung cancer, although it took another 14 years for the US Surgeon General's Report in 1964 to start the long process of educating the public and convincing conservative physicians of tobacco's numerous harms. It is now known that cigarette smoking causes 87 % of lung cancer deaths and that not smoking can significantly reduce not only the risk of other cancers, but also heart attacks, stroke, and chronic lung disease. Once the culprit of a disease is found, it is relatively easier to design therapy that blocks or prevents the evolution of that disease.

In Alzheimer's, the culprit was first thought to be a loss of neurons in the brain that contains the important neurotransmitter acetylcholine (see Chap. 5). Four drugs that increase the amount of acetylcholine by inhibiting the breakdown enzyme acetylcholinesterase in the brain were quickly developed and approved by the FDA for the treatment of Alzheimer's. Three of these drugs still remain on the market. However, it turned out that although acetylcholine was indeed reduced in Alzheimer's, it was a late event and not the cause of this dementia. It is almost charitable to say that all three cholinergic drugs have a very modest effect on reducing Alzheimer's symptoms while their side effects include nausea, vomiting, diarrhea, loss of appetite, and upset stomach.

This void in knowledge was quickly replaced with the amyloid hypothesis. This concept proposed that a gunky material called amyloid beta, or Abeta for short, was

J.C. de la Torre, *Alzheimer's Turning Point*,
DOI 10.1007/978-3-319-34057-9_11

secreted onto the brain tissue by sick neurons or leaky blood vessels and was responsible for everything that happened in Alzheimer's disease. Despite the lack of verifiable evidence in humans, this seductive proposal made the pharmaceutical industry very happy because now they had a therapeutic target to work with, namely, getting rid of the Abeta from the brain. Billions of dollars were poured into the development of a slew of drugs to stimulate the immune system and wipe out Abeta from the brain of Alzheimer's patients. The pharmaceuticals engaged in this venture were salivating at the prospect of cashing-in on a possible mother lode of profits.

None of these "Abeta wipeouts" worked as expected and some drugs even made the patients worse when they developed brain swelling (see Chap. 5). This outcome was an achievement in self-destruction since there is little worse in medicine than acquiring Alzheimer's dementia. Repeated clinical failures have not stopped or made pharmaceuticals rethink the problem from attempting to develop similar Abeta wipeouts (called "me-too" drugs) because as one of their executives was quoted in the Wall Street Journal, "it's a multi-billion dollar market, baby!"

The original strategy by the pharmaceuticals to develop a series of Alzheimer's drugs for patients whose brains were nearly or completely destroyed with no hope of recovery can only be called shameless and morally bankrupt. It was never clear why corporate executives of big pharma with any compassion would wish to extend the misery of a patient, whose memory, intellect, reasoning, and sense of life are all but vanished, qualities that would be replaced with pills promoting confusion, depression, and chronic suffering (see also Chap. 5 for specific details and references).

Why Are Pharmaceuticals Despised?

What makes US pharmaceuticals such a despised industry? Pharmaceuticals are fused to the business school maxim that the only duty of a corporation is "return on investment". This gospel has increased the price of medicines to colossal heights as well as led many of its users into financial ruin. The people hurting the most from high drug prices are the elderly, which is the population group that most need these medicines and the least able to afford them. Many life-saving drugs that cost one-tenth the price in other countries have become the poster child of greed and obscene profit in this nation. This sick corporate culture has even led some **pharmaceuticals to award neuroscience prizes to academic scientists that labor under their banner**, thus cynically increasing the legitimacy of their products.

More disturbing yet, is the jovial cradling by some paid academic consultants who continue to back the dozens of clinically failed anti-Abeta products without challenging the logic or safety of such medical practice. This fawning behavior is reminiscent of an organ grinder's monkey dancing to please its master and get a few peanuts in return.

The pharmaceutical industry justifies these practices by declaring that they spent billions of dollars in their search for new medicines. The irony behind the claim by big pharma that billions of dollars are apportioned to research and development of new drugs is farcical.

Pharmaceuticals often ghost write self-promoting medical articles and pay doctors to slap their names on some medical review or commentary, then publish these paid "advertisements" in prestigious medical journals [1]. Often, human medical trials conducted by pharmaceuticals with negative data are not reported and the results of such trials are not disclosed if they are requested by the FDA or some other source. Most pharmaceuticals pray on the ill-informed, the hypochondriacs, and the sick and injured with expensive remedies that are no better than similar inexpensive ones, renaming old maladies to make them seem more serious than they are in order to boost sales of some medication they have patented.

What about the pharmaceutical claim that they are the main inventors of breakthrough drugs that cost them billions to discover and manufacture? According to **Dr. Marcia Angell** [1], a former editor-in-chief of the distinguished New England Journal of Medicine points out, only a handful of truly innovative drugs have been brought to market in recent years and many of them were developed in academic institutions or small biotechnology companies, usually with the help of taxpayer funding from the National Institutes of Health.

Presently, the pharmaceuticals engaged in "Abeta brain wiping" have developed a new strategy. Probably, as a result of investors prodding, some pharmaceuticals are abandoning their treatment of moderate-to-severe Alzheimer's and are now targeting the disease at a much earlier stage of dementia, where only mild symptoms of memory impairment are present. This is a considerably more intelligent protocol than treating a brain already destroyed by neurodegeneration. But the key issue still remains as to whether wiping out Abeta or preventing its heavy deposition in the brain before symptoms of Alzheimer's blossom will help patients combat the disease in any way. Drug companies such as Pfizer, Elan, Johnson & Johnson, Eli Lilly, to name a few, are not fond of discussing post hoc evidence that their approach to developing anti-Abeta therapy could be on the wrong track since this might suggest to their investors that they don't know what they're doing. There does not appear to be a wake-up call that their greed overwhelms their thinking. Wall Street figures that the pricing of any of these intravenous amyloid-lowering agents, if approved by the FDA, could range from $5,000–20,000 annually per patient; multiply that by the millions of patients who already have or will develop Alzheimer's and it is easy to see that market generating billions of dollars. Pharmaceutical thinking is to not give up on nostrums that have cost them hundreds of millions to clinically test. It is like a horse that cannot win races because it cannot run. Pharmaceutical strategy would be to make the horse run on a shorter track or sell it to suckers if all else fails.

The USA is the only country in the developed world that allows drugmakers to set their own prices. This gift from Congress allows **the ten leading drug companies in the Fortune 500, the combined yearly profits of nearly $36 billion over the profits of all the other 490 businesses put together which total "a mere" $34 billion** [2]. **Not only is the profit margin of drugmakers obscene, many of the medicines they sell are of dubious benefit and some actually have killed thousands of Americans before being removed from the market by the**

FDA. Often, such recalls of harmful drugs comes years after their entry into the market, thus giving pharmaceuticals a temporary passport to make a tidy profit from the misery they created.

Dr. Ben Goldacre is a physician who has criticized the pharmaceutical industry for putting their needs ahead of good medicine. In his book *Bad Pharma* [3], Goldacre writes:

> Drugs are tested by the people who manufacture them, in poorly designed trials, on hopelessly small numbers of weird, unrepresentative patients, and analyzed using techniques that are flawed by design, in such a way that they exaggerate the benefits of treatments. Unsurprisingly, these trials tend to produce results that favour the manufacturer. When trials throw up results that companies don't like, they are perfectly entitled to hide them from doctors and patients, so we only ever see a distorted picture of any drug's true effects. Regulators see most of the trial data, but only from early on in a drug's life, and even then they don't give this data to doctors or patients, or even to other parts of government. This distorted evidence is then communicated and applied in a distorted fashion. In their forty years of practice after leaving medical school, doctors hear about what works ad hoc, from sales reps, colleagues and journals. But those colleagues can be in the pay of drug companies – often undisclosed – and the journals are, too. And so are the patient groups. And finally, academic papers, which everyone thinks of as objective, are often covertly planned and written by people who work directly for the companies, without disclosure. Sometimes whole academic journals are owned outright by one drug company. Aside from all this, for several of the most important and enduring problems in medicine, we have no idea what the best treatment is, because it's not in anyone's financial interest to conduct any trials at all. These are ongoing problems, and although people have claimed to fix many of them, for the most party they have failed; so all of these programs persist, but worse than ever, because now people can pretend that everything is fine after all [3].

It is clear that clinical trials of drugs on humans can represent the difference between life and death, that is, life if done correctly and shown to be effective and safe, and death or damage if not done to meet medical and ethical standard. On this score, Goldacre points out:

> If I were to run a study, and then just remove half of my data points so that my results looked much better, well, you would laugh in my face. It would be obvious to anyone that it was research misconduct. You might even call it fraud. And yet we tolerate the results of entire clinical trials—a huge proportion of them—being withheld from doctors and patients. In medicine, we rely on summaries of evidence and we collate the results from many different trials. So withholding the results of whole trials is exactly the same insult to the data as fraudulently deleting data points from within individual studies.

When Good Medical Research Works

Good medical research should lead to useful clinical trials that can test the merit of a drug in alleviating suffering and disease. Consider the research and clinical trials involved in the development of treatments for HIV/AIDS. This sexually transmitted retroviral infection was a certain death sentence when acquired before 1995. **Dr. David Ho** of the Rockefeller University was one of the early pioneers in the groundbreaking thinking of attacking the HIV virus aggressively before it reproduced itself massively and transformed itself into the deadly AIDS. **This thinking went against expert opinion** that the HIV virus was only dormant in the body and treatment should be withheld until the full-blown AIDS symptoms appeared. By then, as we now know, it was too late.

Ho and his colleagues developed a "cocktail" made up of protease inhibitors and reverse transcriptase inhibitors and treated patients with early HIV. The results of this combination therapy were dramatic in halting the progression to AIDS. While this treatment is not a cure for AIDS, it does show again that **novel thinking, good research, and careful clinical trials can turn around a lethal disease** where life expectancy increased from a few months or short years in an HIV/AIDS patient to a normal life span. HIV treatment today is about **early prevention** and **education of the public**, a rationale that can be helpful in reducing Alzheimer's incidence.

References

1. Wynder EL, Graham EA. Tobacco smoking as a possible etiologic factor in bronchiogenic carcinoma: a study of six hundred and eighty-four proved cases. JAMA. 1950;143:329–36.
2. Angell M. The truth about drug companies. How they deceive us and what to do about it. New York: Random House; 2005.
3. Goldacre B. Bad pharma: how drug companies mislead doctors and harm patients. New York: Faber and Faber, Inc; 2012. p. 100–121.

Chapter 12
Conditions That Can Promote Alzheimer's

Vascular Risk Factors: The Most Important Therapeutic Targets for Alzheimer's Today

The pathogenic link between vascular risk factors and cognitive loss was first suspected from circumstantial animal research that tied insufficient blood supply to the brain with cognitive deterioration. Animal data prompted several large, population-based, prospective studies, including the Kungsholmen, Cache, Rotterdam, and Caerphilly studies. These and other population-based studies reported that vascular risk factors appeared instrumental in the clinical development of Alzheimer's [1].

There is a general agreement that the ability of the brain to maintain normal cognitive function is dependent on optimal cerebral blood flow delivery. The brain is totally dependent on nutrients obtained from the foods we eat to carry out normal neurometabolism, neurotransmission, cognitive tasks, and conscious awareness. Lacking nutrients from insufficient brain blood flow results in a subnormal cerebral state vulnerable to neurodegeneration and dementia.

It is not resolved how vascular risk factors induce cognitive impairment leading to Alzheimer's. However, it became clear to us as early as 1997, based on our animal findings and on key epidemiological reports, that vascular risk factors were critical precursors and a likely trigger of Alzheimer's dementia [2]. Vascular risk factors are now known to contribute to chronic brain hypoperfusion that progressively generates a neuronal energy crisis whose outcome is slowdown of nerve cell metabolism, plaque and tangle deposits, synaptic failure, and brain cell death, mainly in cognitive-related brain regions [1, 2].

It has been estimated that 50 % of Alzheimer patients have modifiable vascular risk factors that could be targeted for treatment with the idea of reducing Alzheimer prevalence [3]. This estimate may be understated because it is unlikely for an older individual reaching age 65 not to have one or, more likely, several risk factors for Alzheimer's. For example, according to the Centers for Disease Control and

J.C. de la Torre, *Alzheimer's Turning Point*,
DOI 10.1007/978-3-319-34057-9_12

(a) BRAIN-RELATED ALZHEIMER RISK FACTORS

Condition	Lowering Effect on Brain Blood Flow
• ischemic stroke	sudden, local blockade of brain vessel(s)
• silent stroke	same as ischemic stroke but asymptomatic
• head injury	chronic brain hypoperfusion
• migraine	abnormal coupling of micro vessels to metabolism
• lower education	possible lack of neuronal activation
• hemorheologic	vessel distortions, red cell viscosity, resistance
• depression	limbic system hypoperfusion hypothesized
• atherosclerosis	obstruction of circle of Willis vessels
• arteriosclerosis	intima media thickening, flow resistance
aging	most common risk factor, non-modifiable

Fig. 12.1 Brain (**a**), heart (**b**), and peripheral related (**c**) risk factors share a common activity: Each factor is a burden to age-related brain blood flow reduction. These "vascular risk factors" have different pathologies, prognoses, and treatments. The risk of Alzheimer's increases exponentially when two or more vascular risk factors are present in the aging individual, ostensibly because more burden on cerebral blood flow results. In this list, only aging is an unmodifiable risk factor. All other risk factors are therapeutic targets for Alzheimer delay or prevention. See text for details. Key: *L* ventricular; *hypt* left ventricular hypertrophy

Prevention, about 66 % of men and women over age 65 have hypertension, 26 % have diabetes type 2, and about 85 % of deaths over age 65 are attributed to atherosclerosis, all of which are modifiable major risk factors for Alzheimer's. In addition, small vessel disease and white matter lesions resulting from exposure to many Alzheimer risk factors in the elderly is a common finding when brain autopsies are performed (Fig. 12.1a–c).

Challengers of the vascular hypothesis of Alzheimer's disease maintain that cerebral hypoperfusion associated with this dementia is due to the neurodegenerative loss of neurons and lower metabolic demand which no longer requires the metabolic energy needed for their survival. However, studies have shown that this conclusion is incorrect. Biochemical and neuroimaging findings have shown that **cortical hypoperfusion in Alzheimer's exceeds the reduction in metabolic demand for oxygen** and is therefore primarily pathological [4]. The reason is that if the brain hypoperfusion were the result of less oxygen requirement needed, the regional oxygen demand would be unchanged or even reduced.

(b) HEART-RELATED ALZHEIMER RISK FACTORS

Condition	**Lowering Effect on Brain Blood Flow**
• heart failure	low cardiac output
• coronary artery disease	hypoxia, low cardiac output
• valvular disease	low cardiac output
• atrial fibrillation	brain ischemic infarcts
• L ventricular hypt	myocardial ischemia, cognitive dysfunction
• hyperhomocysteinemia	possible aortic stiffness, stroke promoter
• hypertension	increased arterial resistance
• hypotension	diminished blood flow volume to brain

(c) PERIPHERAL-RELATED ALZHEIMER RISK FACTORS

Condition	**Lowering Effect on Brain Blood Flow**
• erectile dysfunction	vasoconstriction
• sleep apnea	intermittent hypoxia, cerebral hypoperfusion
• diabetes type 2	vasoconstriction
• hemorheologic	aortic stiffening, basement membrane thickening
• alcoholism	chronic brain hypoperfusion
• menopause	reduced brain blood flow from low estrogen,
• smoking	vasoconstriction
• hyperlipidemia	promotes atherosclerosis and ischemic stroke
• carotid atherosclerosis	vessel blockade
• obesity	associated hypertension, diabetes, sleep apnea

Fig. 12.1 (continued)

In addition, ratio of myelin-associated glycoprotein to proteolipid protein 1 (MAG:PLP1) was measured in Alzheimer brains postmortem and was found significantly reduced in the cerebral cortex [4]. This protein is a marker known to decline when chronic brain hypoperfusion is present. This finding indicates that brain hypoperfusion in Alzheimer's is not simply a physiological adaptation of blood flow that occurs as a result of lower metabolic demand [4].

Other studies have reported that cerebral hypoperfusion is present in cognitively normal, aged individuals who later develop Alzheimer's [5, 6]. These findings suggest that if neurodegeneration were the driving force behind cerebral hypoperfusion, one would expect pathological symptoms to be present prior to such hypoperfusion. Since the opposite seems to be true, it would appear that cerebral hypoperfusion drives neurodegeneration in brain tissue, including the formation of pathological fragments such as Abeta plaques and neurofibrillary tangles.

As we survey the field of Alzheimer research from the present crossroads, it is important to maintain a sense of purpose from the train wreck that characterizes Alzheimer clinical research for the last three decades. A great deal of evidence has been gathered from in vitro and from animal experiments concerning the biochemistry, physiology, and pathology of Alzheimer-related neurodegeneration, but curiously, this information has not translated into a clear clinical advantage to help dementia patients.

However, one bright prospect stands out from the clinical research accumulated thus far. **There is reasonable and consistent evidence that risk factors for Alzheimer's exist and that their control may delay or prevent severe cognitive loss**. This is a pivotal finding that can transform the direction where the Alzheimer field is headed and whether **a turning point** in the management of this dementia is in the making.

The reason for this optimism is that Alzheimer risk factors constitute a presumed link to dementia. This link could be causal. It is not merely that these risk factors "accelerate" Alzheimer's onset, as some investigators now believe; but more importantly, they appear to be direct pathogenic contributors to dementia. This assumption would support the notion that Alzheimer's is not a "unifactorial" disease with one exclusive cause, but rather a heterogenous disorder characterized by an assortment of medical conditions acting to increase the risk of this dementia (Fig. 12.1a–c).

Major findings from recent independent epidemiological studies have identified several dozens of Alzheimer risk factors by examining local Alzheimer populations [7]. The collective data indicate three fundamental findings: (1) Risk factors are potential precursors of Alzheimer's; (2) virtually, all risk factors are associated with reduced cerebral blood flow; and (3) aggregate development of two or more risk factors increases risk for dementia exponentially [2, 7–9]. These observations have received uniform support from multidisciplinary human research involving neuroimaging, psychometric testing, neuropathological examinations, and epidemiological studies [10–12].

It will be recalled from Chap. 1 that as we age, cerebral blood flow is progressively reduced. It is estimated that at age 60, a 15 % reduction of brain blood flow has occurred since age 20, much of it due to cardiac and blood pressure changes. This is generally not a serious problem due to a physiological mechanism called **cerebral autoregulation** which makes sure the amount of blood flow to the brain remains unchanged despite any ups or downs in blood pressure. However, autoregulation is not perfect, and at blood pressures exceeding 160 systolic or lower than 60 systolic, the autoregulatory protection of the brain may be lost. Cerebral autoregulation is also inadequate when chronic brain hypoperfusion occurs. The reason is that cerebral autoregulation works like a bank account. Withdrawal of money can be made to pay creditors, but there reaches a point where there is little or no more money in the account to chronic withdrawal without reserves. Same thing happens with reduced blood flow to the brain, for example, in the presence of chronic brain hypoperfusion. **The key point to discerning how such Alzheimer risk factors impact cognitive function in the aging population is by recognizing their relative effect in reducing cerebral blood flow** [2, 13, 14, 15]. The reduced blood flow in the elderly brain is age-related, but an additional burden can be added to this blood flow insufficiency when the individual develops one or more risk factors.

Here is how it works. A look at Fig. 12.1a–c shows that despite the distinct pathologies and etiologies associated with each of these risk factors as well as their differential clinical course and prognosis, **all share a common action: the reduction or impairment of cerebral blood flow**. This fact alone provides a unique insight into the cause, natural history and pathologic process involved in Alzheimer's disease and opens fruitful pathways to prevent the evolution of dementia. Because of the extraordinary relationship between Alzheimer risk factors and medical conditions that reduce cerebral blood flow, we shall, from this point onward, refer to these conditions as **vascular risk factors for Alzheimer's disease**. The vascular risk factors listed in Fig. 12.1a–c contribute to lowering brain blood flow in a variety of ways. Let us review some of these ways.

Brain-Related Vascular Risk Factors

Brain-related risk factors for Alzheimer's disease (Fig. 12.1a) can reduce cerebral perfusion through blockade of the vessel lumen from the long-term buildup of atherosclerotic plaques or from a sudden, ischemic stroke forming a clot within the vessel wall which can be symptomatic or asymptomatic (silent). Ischemic stroke affects about 700,000 people in the USA every year of which about 130,000 die. Many survivors require long-term physical and occupational therapy to cope with their neurological deficits. It is estimated by the American Heart Association that about 43 % of the elderly individuals in the USA harbor an asymptomatic, silent stroke, but it is not known how many of these elderly individuals develop cognitive loss as a result.

Physical narrowing of the vessel lumen (vasoconstriction) is another common way that chronic brain hypoperfusion can develop from brain-related risk factors [2]. Blood vessel vasoconstriction or vasodilation in the brain is a complex, not well-understood, and often paradoxical dynamic process involving vasoactive substances released by astrocytes whose endfeet make contact with the outside wall of cerebral arterioles or by endothelial cells located on the inner microvessel wall. The complexity of human cerebral blood flow regulation becomes somewhat palpable when it is estimated that a single astrocyte with its many processes can regulate over a million synapses in its microenvironment and only one of its processes with many endfeet is needed to make contact with a short length of an arteriole or capillary to regulate vessel blood flow.

The interaction (called signaling) between brain endothelial cells and astrocyte endfeet in the dynamic control of arteriolar tone is not clearly understood. Tone is the ability of an artery or arteriole to change its diameter using the vessel's smooth muscle cells on its outer wall in response to a pressure stimulus from within the vessel. This classic phenomenon is known as the "Bayliss effect" and is a fundamental contributor to cerebral autoregulation.

Release of vasoactive molecules to control vessel tone is probably initiated by commands from neurons to astrocyte endfeet and to endothelial cells that control nitric oxide in an attempt to maintain an adequate perfusion rate in the face of blood pressure changes, sudden blood loss, cardiac arrhythmias, or some other pathological event that threatens neuronal survival. **Compensatory mechanisms for faulty cerebral blood flow do not always work as described above, particularly where chronic brain hypoperfusion is involved**. The reason is that although there is a vasoactive network of adaptive mechanisms to cope with short-term falls in blood flow delivery to brain cells, chronic brain hypoperfusion may overcome these compensatory mechanisms due to its persistent and unrelenting nature. **I believe that if the source of cerebral hypoperfusion can be found and controlled, for example, from conditions listed in** Fig. 12.1a–c, **progressive cognitive deterioration may be prevented or significantly delayed**. This important issue is discussed in Chap. 16.

One of my recent reviews on the topic of cerebral hemodynamics and blood flow gives a crystallized summary of this process using a minimum of technical language [15]. **What seems important in attempting to understand the behavior of blood flow in the brain is that it provides important clues in seeking ways to prevent or control abnormal blood flow patterns initiated by aging coupled to vascular risk factors**. These abnormal blood flow patterns are the basis for **the vascular hypothesis of Alzheimer's disease** that we first proposed in 1993 [14].

For example, age-related blood vessel narrowing that induces mild cerebral hypoperfusion can be detected during middle age but more so after age 60 when 15–30 % of normal blood flow may be decreased for years or decades without any appreciable dyscognitive symptoms being manifested except for occasional memory lapses. When mild symptoms of age-related cerebral hypoperfusion occur in the presence of one or more Alzheimer risk factors, a condition may develop that we have called **critically attained threshold of cerebral hypoperfusion (CATCH)** [15]. CATCH may explain the so-called senior moments when an elderly individual occasionally forgets a word, an appointment, a place, or an event. Senior moments commonly represent absentmindedness or a temporary distraction. It may also be a precursor to mild cognitive impairment or represent a red flag where more serious memory problems have yet to come. More about this is in Chap. 17. Mild cognitive impairment is not a straight path to dementia since most individuals diagnosed with this condition never develop Alzheimer's and never progress beyond a mild stage of cognitive difficulties.

A neuroimaging test called FDG-PET (fluorodeoxyglucose positron emission tomography) has been used to predict conversion to Alzheimer's in mild cognitive impaired patients [16]. I do not recommend anyone taking this test simply to obtain a result that may be positive or negative in predicting Alzheimer's because it only introduces stress and anxiety for the patient in the event the test is positive. Also, since the test is not close to being 100 % accurate in its prediction, the negative or positive result may offer either a false sense of security or dreaded anguish about a lurking disease that may never materialize. The exception to my recommendation is discussed in Chap. 16 where a strategy for the management or treatment of detected vascular risk factors for Alzheimer's disease is coupled to **Heart-Brain Clinics** devoted to the prevention of cognitive deterioration.

Ischemic and silent strokes are associated with the eventual development of vascular dementia affecting mostly the elderly, but they are also seen prior to Alzheimer's. People who have an ischemic stroke more than double their chance to develop Alzheimer dementia. There is no clear explanation so far for the link between a sudden brain ischemic event and the advent of Alzheimer's. Suspicious co-conspirators also called comorbid conditions such as hypertension, obesity, cardiovascular anomalies, atrial fibrillation, sleep apnea, certain medicines, and blood disorders that promote blood viscosity and increased coagulation could accelerate the formation of large arterial or small blood vessel strokes. The fact that cognitive symptoms following an ischemic stroke are virtually identical in vascular dementia and Alzheimer's indicates that these two dementias overlap considerably and may differ only in the time it takes to express the symptoms characteristic of either dementia.

The region of the brain affected by a stroke can influence the potential outlook to dementia. Strokes that affect the left side of the brain have a greater risk of dementia than those affecting the right side, probably because the left hemisphere is associated with more cognitive functions than the right hemisphere.

One vascular event that has received little research attention in relation to its clinical gravity is the development of **silent stroke**. It has been estimated that approximately 11 million Americans experience a silent stroke every year, defined as a focal stroke without acute symptoms [17]. Silent stroke shows a higher prevalence in cigarette smokers and people with atherosclerosis. Smoking and atherosclerosis are conditions linked to Alzheimer's and to cerebral hypoperfusion. Silent stroke may be a "sleeping giant" in the development of Alzheimer's since cerebral perfusion is often found to be reduced in association with "misery perfusion" during an attack, a dynamic blood flow deficit typically found in Alzheimer patients.

Head Injury

Head injury was one of the first risk factors to be reported for Alzheimer's disease. The increased risk of Alzheimer's after a traumatic brain injury is more commonly seen when loss of consciousness occurs. However, not all severe head injuries show a link to dementia, an observation that has not been fully understood. Symptoms of cognitive dysfunction may follow soon after a head injury but can also lie dormant for decades and blossom into mild to severe cognitive signs of impairment. This is particularly seen when repeated concussions are sustained in athletes playing football, hockey, soccer, or other sports where the head is a target for concussive injuries. The fast rise of chronic traumatic encephalopathies (CTE) recorded today among professional football players is an example of big money interests ignoring a problem that has existed for many decades and is only being recognized through public complaints made by well-known former football stars. There is no clear consensus on the quantity or duration of concussions needed to cause CTE. The symptoms of CTE initially begin with poor concentration, attention deficits, and memory lapses along with dizzy spells and headaches. These symptoms worsen over time affecting social functioning, impaired memory, poor decision making, social phobias, paranoid ideation, flawed relationships, insomnia, and unpredictable mood swings [18]. The clinical picture resembles that of vascular or Alzheimer dementia.

Dementia pugilistica that results in progressive neurological and cognitive deterioration in boxers who sustain repeated blows to the head has been known since the 1920s and dismissed as an unfortunate consequence of an entertaining sport. Over years or decades, these repetitive insults to the brain are characterized by progressive atrophic changes in the medial temporal lobe, thalamus, mammillary bodies, and brain stem; these lesions are accompanied by ventricular enlargement capable of compressing the brain tissue and creating additional damage [19, 20] (Fig. 12.2).

It would seem that a single hard blow to the head or repetitive milder concussions will reduce cerebral blood flow, especially during advanced aging to a point where cognitive deterioration and neurological sensorimotor deficits will result. The physiological explanation for how physical brain trauma reduces blood flow is lacking, but evidence has indicated that lower perfusion pressure from a compromised microcirculation may be a major contributor to this outcome [19]. Studies of people who had a history of a previous traumatic brain injury have shown that global hypoperfusion is detectable in their posterior cingulate cortex and medial frontal lobe cortex; these are brain regions which have been shown to be hypoperfused in Alzheimer patients with no history of trauma to their brain [5, 21].

These findings suggest that the impact of traumatic brain injury or repeated brain concussions can lie symptomless for many years during a period of cerebral hypoperfusion and then, for unknown reasons, begin to express a progressive cognitive decline not unlike the clinical pattern seen in patients with senior moments who develop mild cognitive impairment and eventually, Alzheimer's disease.

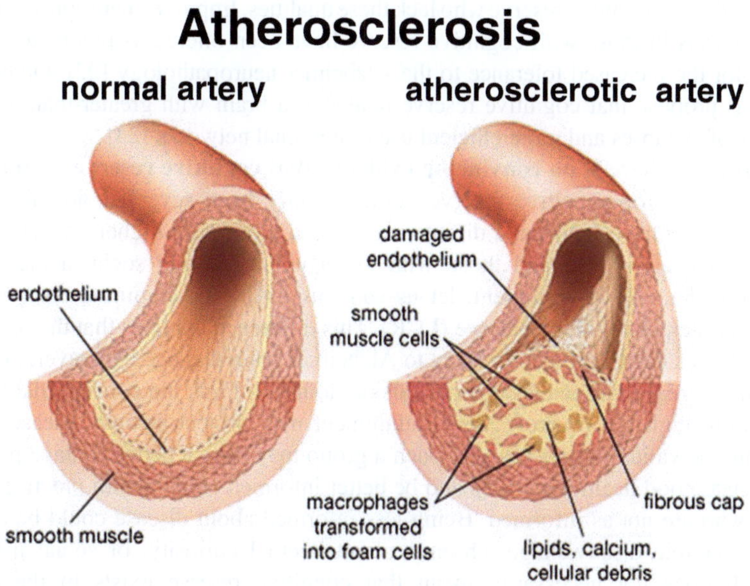

Fig. 12.2 Cross section of an artery affected by atherosclerosis. Compared to a normal artery, clumping of blood cells, narrowed lumen, and damage to endothelial cells which line the inside wall of the vessel are seen. Clumping of blood cells, debris from vascular smooth muscle, and macrophages contribute to the complex formation of a mature atherosclerotic plaque. Endothelial damage in smaller arteries and arterioles can prevent the release of vasoactive substances which can impair the artery's ability to provide cerebral autoregulation when blood pressure changes occur. Smooth muscle cells, whose role is to maintain vessel tone, can accumulate and proliferate in the arterial wall after atherosclerosis. Symptoms occur when there is sufficient obstruction of the lumen to cause ischemia to the end organ, including brain. Plaques can also rupture sending emboli to occlude up- or downstream organs. In the case of the brain, plaque rupture can contribute to a silent or active stroke. See text for more details

Other Common Alzheimer Vascular Risk Factors

Other brain-related risk factors for Alzheimer's such as **migraine**, **lower education**, and **depression** are less well understood with respect to their vascular involvement and how they are able to lower cerebral perfusion. Findings suggest that at least some migraine patients show lesions in the brain compatible with a previous white matter infarcts and that these patients have a higher risk for cognitive impairment; however, these findings have been debated and challenged.

Lower education as an Alzheimer's risk has been studied intensely. Although the link between lower education and cognitive deterioration appears to exist, the reason for the link is unclear.

Some have suggested **cognitive reserve**, a hypothetical construct defined as people with a superior IQ, professional accomplishment, and educational attainment or enriched environment, which are better able to cope with Alzheimer-related damage than those with less cognitive reserve who lack these qualities. Improved neuroplasticity and neurogenesis in those with cognitive reserve have been offered as potential mechanisms for the increased tolerance to the Alzheimer neuropathology [22]. Other suggestions propose that cognitive reserve relates to a brain with greater-than-average density of synapses and more efficient use of neuronal networks [23].

However, there is no convincing evidence that cognitive reserve even exists. The evidence for cognitive reserve comes mainly from several epidemiological studies which have reported a difference between those with higher education and professional attainment and those with lesser education and social achievement [24]. For the sake of argument, let us call this difference cognitive reserve plus (CRP) or lack of cognitive reserve (LCR). This difference assumes that the presence of CRP has a causal protective effect to Alzheimer brain damage. However, an error in *post hoc* reasoning makes this conclusion doubtful. First, the concept that higher education, high IQ, or professional attainment may protect against disease could certainly be valid but only because such a group may have a better vantage point to appreciate good health measures and be better informed about health practices than those who are not as informed. Being less informed about disease could be due to socioeconomic disadvantage, absence of intellectual curiosity, or social indifference. It does not necessarily mean that cognitive reserve exists in the better informed group since other factors may be responsible for the CRP:LCR difference. The reason is that no one to our knowledge has ever proven that the brains of people who test positive for CRP are endowed with richer neuroanatomic, neuroplastic, or neurophysiological traits than those with LCR.

We have long maintained that what seems to be important in the resilience of the brain to mildly protect against age-related wear and tear damage is not the degree of higher education, innate intelligence, or occupational attainment but the ability to engage in vigorous mental activities that induce a boost in cerebral blood flow [25]. A boost in daily cerebral blood flow can be achieved with a host of mental activities [26, 27]. These mental activities are more likely to be used by people with higher education, intellectual achievement, and professional attainment but are not exclusive to this group. Excluding the high-Alzheimer-risk $APOE_4$ genotype from the formula, a low-educated or blue-collar worker with intellectual curiosity can engage in critical thinking and intellectual endeavors to achieve the same degree of protection from age-related cognitive decline as CRP persons. Ernest Hemingway never went to college and William Faulkner did not finish high school; both won the Nobel Prize in Literature.

The protective effects of mentally stimulating activities on cognitive impairment has been demonstrated by a recent study by Wilson and colleagues [28] who recruited over 1000 cognitively normal elderly individuals with a mean age of 80 and **diverse levels of education and professional attainments**. It was found after a 5-year follow-up that those individuals who engaged in sustained, mentally stimulating activities such as playing chess, reading newspapers, and visiting a library showed better cognitive function than those who did not engage in such mental activities [28]. These findings implicate the generally accepted correlation between neurovascular coupling and cognitive function where engaging in repetitive mental activity increases metabolic demand and consequently requires a higher supply of cerebral blood flow. The resultant average increase in cerebral blood flow following mental reasoning activities appears to be modest, about 10 % from baseline [27], and may be short-lived [29]. But if performed on a routine basis, it may explain what is believed to be cognitive reserve in socially priviliged individuals who are more prone to engage in such mental activities.

Following this reasoning, studies have shown that mental activity requiring certain types of complex thinking can increase oxidative metabolism and regional cerebral blood flow in multiple cortical and subcortical fields including the hippocampus, one of the initial sites of Alzheimer pathology [27].

Considering these findings, it is not surprising that participating in stimulating mental activities during advanced aging has been consistently shown to provide a measurable degree of protection from cognitive decline in cognitively intact individuals, regardless of professional status, IQ, or educational level. Such mental exercises can be compared to the benefits of aerobic physical exercise that have been shown in a meta-analysis review of the subject to improve cognitive function in a number of cognitive domains including executive function [30]. Executive function is an umbrella term that refers to the control and management of a number of cognitive processes including working memory, planning, problem solving, and mental flexibility. Impairment of executive function begins early after a diagnosis of Alzheimer's disease. Since physical exercise is reported to slow down cognitive decline, one must suspect that the reason could be in keeping a healthier heart that can supply the needed blood flow by the brain in time of trouble. This conclusion is speculative but is logically realistic as will be seen below in our discussion of cardiac-related risk factors for Alzheimer's.

Studies of bilingual or multilingual people have also shown that knowing two or more languages offers cognitive and memory processing benefits that could fend off cognitive decline during aging. Again, this cognitive protection seems independent of CRP or LCR.

My inference is that all things being equal, an uneducated but mentally active individual has the same opportunity of lessening the risk of Alzheimer's as a graduate from Oxford employed as a cryptanalyst. Further research is needed to ascertain whether this conclusion has any merit.

Depression, **apathy, and anxiety** may be the first signs of Alzheimer's onset even before memory loss is observed. Excluding apathy and anxiety, depressive symptoms in Alzheimer's appear to correlate with **decreased cerebral blood flow**. Several studies in the mid-1990s actually reported an **increase** in cerebral blood flow in depressive people and challenged the findings that hypoperfusion developed in specific brain regions. These contrasting results may have been due to semiquantitative blood flow measuring techniques using radiopharmaceutical xenon-133, PET scans, or prototype imaging methods. In recent years, functional neuroimaging techniques have improved considerably and provided quantitative results consistent with changes in the hemodynamics of the frontal–prefrontal cortex, limbic system, and temporal subcortical regions in those with depression [31].

A common hypoperfused brain area affected by depression is the left middle prefrontal cortex which is an area presumed to be pathogenetically involved in depressive symptoms [32]. Recovery from depression also correlates with an increase in cerebral perfusion in the left middle prefrontal cortex [32], a finding that indicates that interventions aimed at improving blood flow to the brain may help reverse depressive symptoms. One reason why depression correlates with reduced blood flow is the finding of increased blood viscosity in depressed patients [32]. Blood viscosity is a measure of the thickness and stickiness of blood. Its increase is dependent on two factors: an elevated hematocrit and a high rigidity of red blood cells, a condition where red blood cells have difficulty bending their fluid lipid bilayer to flow easily through a tight capillary lumen. In a normal state, red blood cells take the shape of a bullet from a spherical shape to pass quickly through a tight capillary. Since red blood cells carry oxygen to tissue, their inability to flow freely through capillaries will rob cells from an essential nutrient needed to survive. There are many disorders that will compromise red blood cell deformability in brain capillaries including ischemia and hypoperfusion.

It is tempting to hypothesize whether red blood cell deformability is negatively affected by vascular risk factors capable of inducing brain hypoperfusion and increasing blood viscosity.

Having a high blood viscosity is thought to be linked with heart problems because viscous blood makes it harder for the heart to pump blood and sticky blood cells tend to form clots in coronary and cerebral arteries. We have reviewed the hemodynamics of high blood viscosity with respect to its possible effects in Alzheimer's disease [14, 15]. Hemorheology, the study of how blood flows in the circulation, especially during aging, is not a topic that has attracted much research despite its importance in cardiovascular and cerebrovascular events and their influence on neurometabolism and cognitive function. Indeed, few studies have examined the hemorheology of red blood cell aggregation, fibrinogen levels, and whole-blood viscosity during aging.

Heart-Related Risk Factors

Atherosclerosis

Atherosclerosis can affect the inside wall of arteries by forming an atherosclerotic plaque made up of blood cells, cellular debris, macrophages, and mobile smooth muscle cells [33] (Fig. 12.2). Formation of atherosclerotic plaques will reduce blood flow partially or totally in any artery of the body. The development of atherosclerosis can compromise the supply of oxygen to the brain indirectly through the carotid arteries or directly by obstruction of blood flow in the circle of Willis arteries. An after-death study on patients who had been cognitively normal and those with Alzheimer's was done by comparing the amount of atherosclerosis in the arteries of the circle of Willis [34] (Fig. 12.3). The circle of Willis is a polygon-shaped arterial loop that connects the major arteries supplying the forebrain and hindbrain, which are composed of the internal carotid arteries and the vertebral arteries.

The arterial polygon lies at the base of the brain branches into smaller arteries and capillaries that provide oxygenation and vital nutrients to maintain brain cell

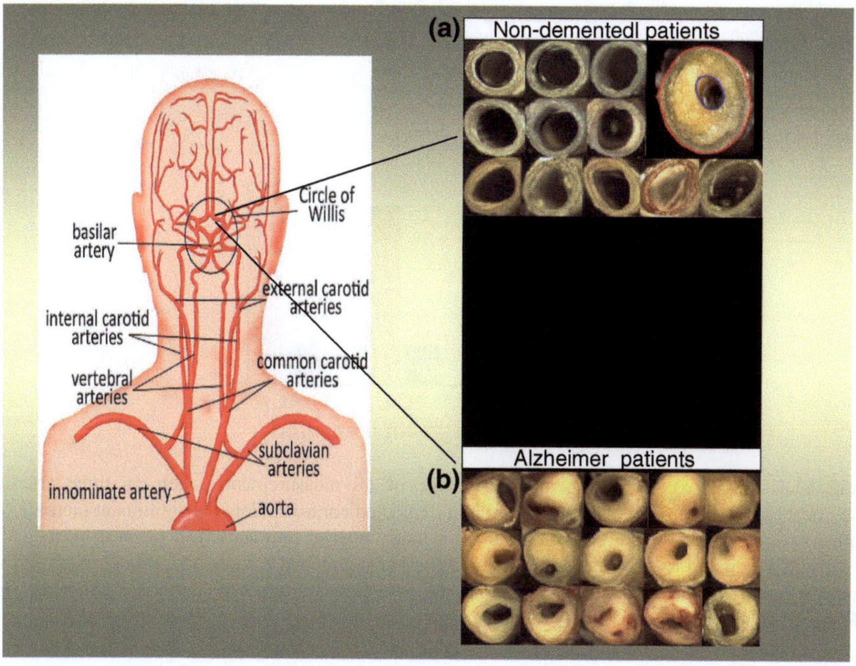

Fig. 12.3 Heart-to-brain circulation. Postmortem cross sections of non-demented (**a**) and Alzheimer's (**b**) circle of Willis arteries. Non-demented circle of Willis arteries (**a**) shows minor narrowing of the lumen compared to severe lumen narrowing (**b**) in Alzheimer circle of Willis arteries indicating marked atherosclerosis. This narrowed pattern in circle of Willis arteries seen in Alzheimer brains was also observed in the arteries that supply blood flow to the brain, namely internal carotids and vertebral arteries. About half of the Alzheimer arteries shown above were 60 % or more occluded. Modified from Roher et al. [34]

survival. Severe narrowing of the lumen from marked atherosclerosis, often by more than 80 %, was observed in the circle of Willis arteries in patients who had died with Alzheimer's disease by Roher and his colleagues [34] (Fig. 12.4). By contrast, only minor narrowing of the lumen was generally seen in cognitively normal patients who had died of other causes [34] (Fig. 12.4). The dramatic findings from this study show convincingly that atherosclerosis is a major culprit in the development and progression of Alzheimer's disease. The study also underlines the importance of lifestyle in preventing or slowing down the formation of arterial atherosclerosis by keeping a healthy diet, lowering fat intake, controlling blood pressure and diabetes type 2, not smoking, and exercising daily.

Atherosclerosis can also invade coronary arteries exposing the heart to coronary heart disease which can lead to chest pain, heart attack, or death. Cerebroarterial and coronary artery disease are two major risk factors for cognitive decline in the elderly that often results in dementia.

Atherosclerosis can attack arteries in the periphery, a common condition in the elderly called peripheral arterial disease. Until recently, peripheral arterial disease

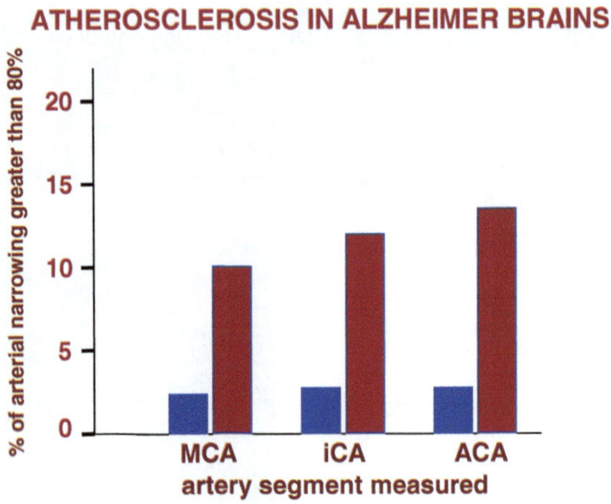

Fig. 12.4 The y-axis represents the percentage of all measurements taken on three arterial segments in the brains of Alzheimer's and non-demented controls. Note the significant increase of atherosclerotic narrowing greater than 80 % in major arterial segments (*red bars*) in Alzheimer brains as compared to non-demented brains (*blue bars*) matched for age and gender. Key: *MCA* middle cerebral arteries; *ICA* internal carotid arteries; *ACA* anterior cerebral arteries. Adapted from Roher et al. [34]

was not known to affect cognitive function, but patient-based studies have shown that individuals with arterial plaques outside the brain and heart tend to perform more poorly on cognitive tests than control subjects [35]. A quick, noninvasive clinic test called ankle–brachial index is used to diagnose atherosclerosis in peripheral vessels and may predict people at risk of developing cognitive impairment [36]. The test compares the systolic blood pressure measured at the ankle with that measured at the arm. If the pressure in the ankle is lower than that in the arm, a blocked artery and peripheral arterial disease is likely to be the cause (Fig. 12.5).

While atherosclerosis causes its damage mostly on the inside wall of an artery, arteriosclerosis acts to stiffen and thicken the muscular walls of arteries. Arterial stiffness is the principal factor that prevents arterial cushioning of blood flow from slamming into the inner walls of large arteries where, in time, elasticity of these arteries is severely lessened. When that happens, the normal hemodynamic forces that carry blood to the brain become like microscopic tornadoes that can damage arterioles and capillaries in the brain to the extent that cerebral blood flow becomes deficient. This hemodynamic process can lead to pathological *shear stress*. Shear stress is the force of the blood applied to the vessel wall. This force can impact the endothelium where vessel diameter is regulated depending on the vessel wall elasticity (distensibility) and on the release of vasoactive agents released by endothelial cells that control vasodilation and vasoconstriction. Abnormal shear stress is an important mechanism in the development of atherogenesis and in arteriosclerosis.

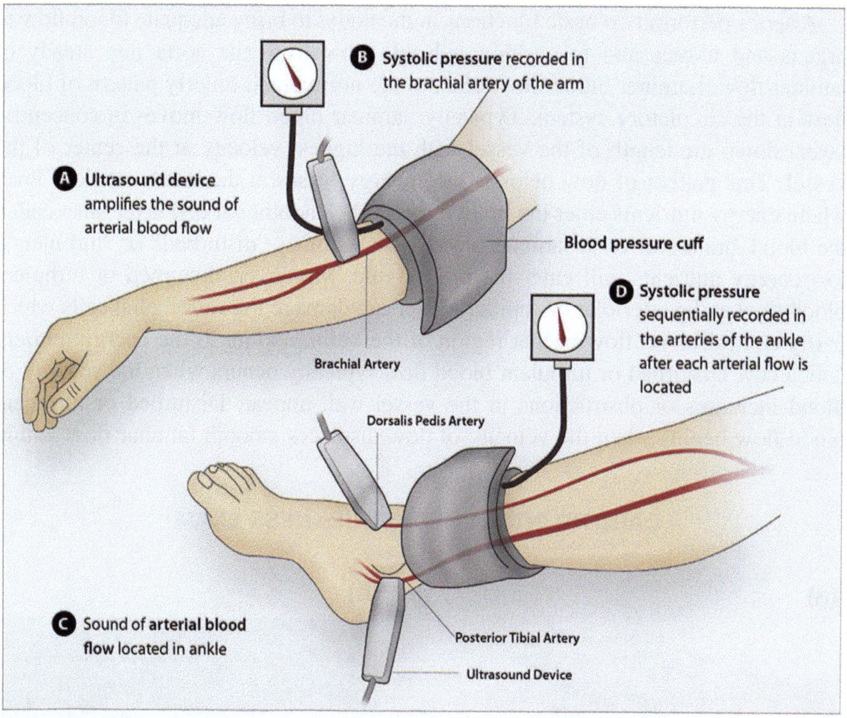

Fig. 12.5 The ankle–brachial test is a diagnostic technique used to detect peripheral arterial disease (PAD) and atherosclerosis of peripheral arteries. The test is positive when the blood pressure in the arm reads higher than that in the ankle. The test is noninvasive and quick to perform and provides useful information that may help reverse or delay cognitive decline in the elderly

The increasing thickness of the muscular wall can bulge toward the lumen where blood flow can be slowed down creating a condition called **intima–media thickening**. Intima–media thickening is commonly observed during advanced aging. It often progresses to atherosclerosis and all its complications. This process creates a **double-hit** to the artery affected by narrowing the lumen physically from a bulging intima–media thickness and by allowing more fat-laden plaques to deposit on the inner vessel wall. When the double-hit strikes the common carotid arteries that supply blood flow to the brain, the heart will need to work harder to overcome vessel resistance and partial blocking of blood flow, a process that lowers the delivery of high-energy nutrients such as glucose and oxygen to the brain. This process is reported to increase the risk of cognitive decline that can lead to Alzheimer's due to insufficient blood flow from reaching brain cells to meet their metabolic demands. To adequately meet these metabolic demands from neurons and glial cells, cerebral perfusion requires about 750 ml of blood flow per minute [37].

Arteries perform two basic functions in the body: to bring adequate blood flow to organs and tissues and to cushion pulsatile flow from the aorta into steady or laminar flow. Laminar blood flow refers to the normal and orderly pattern of blood flow in the circulatory system. Typically, laminar blood flow moves in concentric layers down the length of the vessel with the highest velocity at the center of the vessel. This pattern of flow helps reduce energy losses at the brain capillary level where energy nutrients enter the brain through the endothelial cell layer, also called the blood–brain barrier. If laminar blood flow becomes "disturbed" or "turbulent," low-energy nutrients will enter the brain tissue. Moreover, disturbed or turbulent blood flow at the arteriolar or capillary level can damage the endothelial cells which partly control blood flow in that region of the brain, adding to the energy nutrient deficiency. Disturbed or turbulent blood flow typically occurs when the velocity of blood increases or obstructions in the vessel wall appear. Disturbed or turbulent blood flow begins when the velocity of flow displaces smooth laminar flow within

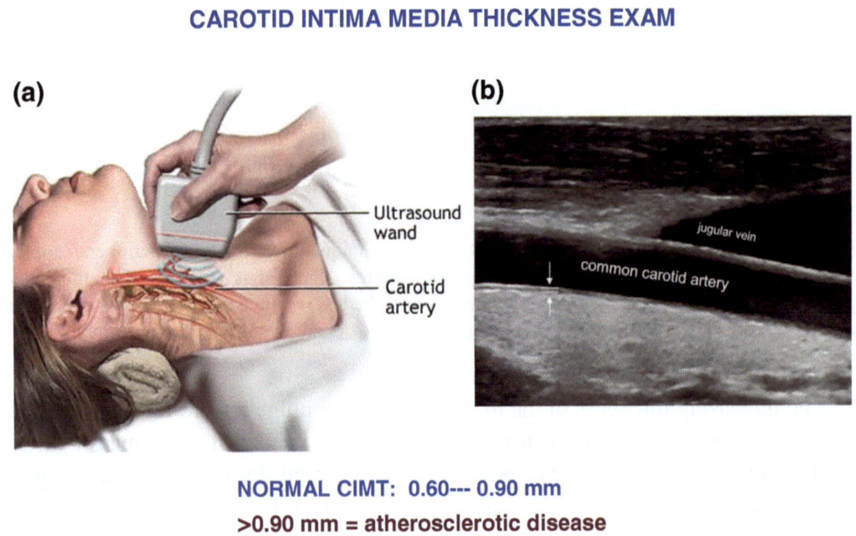

Fig. 12.6 The carotid intima–media thickness (CIMT) is a noninvasive, painless test that examines the thickness of the intimal and medial layers of the carotid artery walls. **a** CIMT is an excellent marker for accurate risk stratification in asymptomatic patients that may be at risk of Alzheimer's. It can detect soft and calcific plaques within the carotid artery and correlates well with plaques in the coronary arteries of the heart. CIMT can also assess blood flow velocity to the brain and can track the usefulness of a therapeutic intervention. The normal combined thickness of intima–media layers ranges from 0.60 to 0.90 mm and tends to increase with age. A thickness greater than 0.90 mm indicates atherosclerosis and risk of stroke and heart attack. Actual high-resolution B-mode ultrasound of common carotid artery wall segments **b** showing region of interest (*arrows*). The region between the two parallel lines corresponds to the lumen–intima interface and the media–adventitia interface reflecting the intima–media thickness. Velocity and direction of blood flow can also be measured when B-mode is combined with pulse wave Doppler scanning. Common carotid artery lumen in healthy older individuals measures 6–8 mm in diameter, and normal intima–media thickness ranges from 0.4 mm at birth to 0.8 mm at age 80

the vessel. These abnormal blood flow patterns can be seen in a number of disorders that result in arterial narrowing or stiffening (see below, **Arterial stiffening**).

Carotid artery intima–media thickness is a marker of atherosclerosis and cardiac disease, and when it is detected using noninvasive Doppler carotid artery ultrasonography (Fig. 12.6), it predicts that cognitive impairment may follow, especially in the asymptomatic elderly person. The test measures the thickness of the inner two layers of the carotid artery—the intima and media. When a specific thickness of this layer is detected in the asymptomatic patient, the physician may opt to initiate aggressive medical treatment or surgery to prevent a possible stroke or reduced perfusion to the brain. As a rule of thumb, when plaque-like material or intima–media thickness is found blocking the carotid artery, coronary artery disease is almost always present.

The primary contributors to carotid artery intima–media thickness are age and hypertension, but not far behind are Western society's contribution to unsuccessful aging, the 3 Ds: diabetes, dyslipidemia, and deficient dieting.

It is not surprising that carotid artery intima–media thickness appears to reflect the aging vasculature since loss of elasticity in the arterial tree tends to thicken 2 to 3 times between ages 20 and 90 while diminishing the ability of the artery to recoil back to its original dimensions following blood flow pressure waves in the vessel. The scientific term applied to this phenomenon is loss of compliance resulting from age-related arterial wall stiffening.

The effects of carotid artery intima–media thickness on cerebral hemodynamic abnormalities are a topic that requires considerably more research attention because if cerebral perfusion can be kept adequate during advanced aging, when it is most vulnerable to hypoperfusion, prevention of progressive cognitive loss and its transition to Alzheimer's or vascular dementia is a realistic goal.

Aortic Stiffening

Considerable evidence now indicates that cognitive impairment can follow chronic aortic stiffness in older individuals by damaging brain microvessels [38]. The aortic valve normally opens to allow blood to pass from the left ventricle to the aorta, the massive blood vessel that directs oxygenated blood from the heart to the rest of the body. When the left ventricle or aortic valve malfunction or aortic stiffness is present, the initial organ damage starts at the brain due to the high demand of this organ for oxygenated blood. The damage to brain microvessels from aortic stiffening would occur as follows. The aorta is known to be a reservoir of pulsatile (beat to beat) energy delivered by the left ventricle pumping blood out. Because the aorta is normally more distensible than the stiffer carotid arteries, it can absorb the ventricular ejection of blood and cushion the pulsatile flow, so it does not proceed full strength into the carotid arteries that supply the brain vasculature. This hemodynamic phenomenon is called the **Windkessel effect**.

The Windkessel effect acts to dampen excessive transmission of pulsatile flow that can damage the cerebral microvasculature and impair astrocyte signaling from neurons that control cerebral blood flow. The normal aorta thus reduces excessive flow force to the carotid arteries and small vessels in the brain, thus avoiding forceful transmission of that flow to the cerebral arterioles and capillaries. The **Windkessel effect is considered a protective mechanism that helps coronary arteries and organs maintain good perfusion**.

However, during aging, there is an increased risk of hypertension, atherosclerosis, and the production of inflammatory proteins. These pathological changes can lead to a loss of elastic lamellae which provide distensibility to the aorta (Fig. 12.7). The outcome of these changes markedly reduces aortic compliance with the net

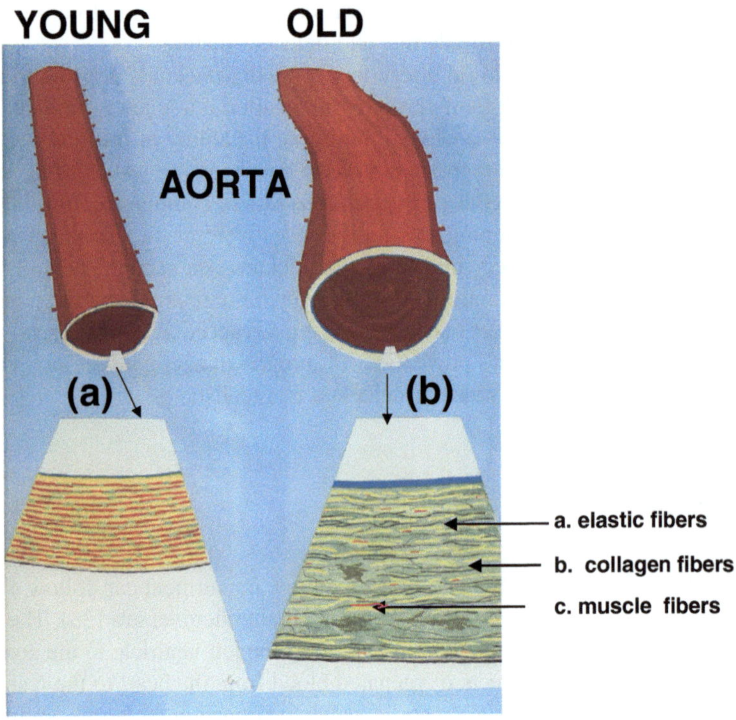

Fig. 12.7 The wall of the older human aorta (**b**) is disorganized as compared to young (**a**) aorta. This is a consequence of fraying and fracture of the elastic fibers (*a yellow*) and loss of muscle attachments (*c red*), together with the increase in collagen fibers (*b black*) and mucoid material (*green*). The clinical outcome of aortic stiffness is increased systolic blood pressure and reduced diastolic pressure. The physiological outcome is reduced ability by the aorta to expand and accommodate to the volume of blood being pumped out of the left ventricle. This inability of the aorta to expand significantly increases pulsatile flow in the brain microcirculation giving rise to structural brain lesions which are associated with lower cognitive function. Increased pulsatile flow may also distort brain microvessels, thus reducing high-energy nutrient delivery to brain cells. See Chap. 10 for details

effect of increasing pulse pressure and systolic blood pressure while reducing diastolic blood pressure and wave reflection at the carotid arteries. The consequences that can occur when distensibility of the aorta is deficient include greater chance for heart attacks, cardiac enlargement, heart failure, stroke, peripheral vascular disease, and renal damage.

Collagen deposition in the heart muscle fibers also contributes to aortic stiffening and reduced dispensability (Fig. 12.7). These hemodynamic abnormalities are worsened if vascular risk factors develop (Fig. 12.1a–c). When this happens, exaggerated beat-to-beat flow and pulse wave velocity (the rate at which pressure waves move through a vessel) are transmitted to microvessels in the brain. Microvessel damage in the brain can result in white matter lesions and damage to arteriolar endothelial cells that participate in controlling cerebral blood flow. **This pathological process suggests a mechanical link between brain hypoperfusion and cognitive impairment whose primary trigger is loss of the Windkessel effect**. The importance of progressive increases of pulse wave velocity in elderly people has been postulated to result in increasing cognitive decline and eventual Alzheimer dementia.

Experimental findings suggest that arterial stiffness and increased pulse wave velocity during aging may occur from a mitochondrial disbalance between healthy production of mitochondria and degradation of damaged mitochondria, a process known as **mitophagy**. If mitophagy can be enhanced in age-related arterial stiffening, it is speculated that the mitochondrial stress and dysfunction resulting from the disbalance between mitochondrial production and degradation could be reversed or managed.

Trehalose, a nutraceutical that enhances mitophagy, was reported to reverse age-related arterial stiffening of the aorta and pulse wave velocity when given as a food supplement to old mice [39]. This study inferred and supported the notion that mitochondrial stress and dysfunction during aging underlie age-related aortic stiffening, possibly in the presence of oxidative stress and inflammatory cytokines. Thus, trehalose and similar non-reducing disaccharides could lower free radical production and pulse wave velocity while normalizing the overproduced aortic collagen I, a key contributor to age-associated arterial stiffness (Fig. 12.7). Downstream, trehalose also appears to act as an antioxidant and free radical scavenger.

Since trehalose is a natural sugar with no apparent serious side effects and can be assimilated much like sucrose, it would be of much interest if human studies could confirm its effect on mice, given that aortic stiffening and pulse wave velocity show marked improvement when rodents are fed trehalose. Improvement of aortic stiffening would be reflected by a slowing or prevention of cognitive decline.

Aortic stiffness has been classically treated with antihypertensive medication. Long-term trials have suggested that angiotensin-converting enzyme inhibitors (ACEIs) are useful in markedly reducing pulse wave velocity. Other blood pressure medications called calcium channel blockers and beta-blockers also have this ability but to a lesser extent.

It seems a wise practice in evaluating elderly patients at cardiac clinics to look for signs of increased pulse wave velocity and peripheral artery disease using ankle-brachial blood pressure test (called ABI), a relatively simple office test and strong predictor of cardiovascular pathology and aortic stiffness.

This test is not generally performed on patients who present cardiac signs and advanced aging due to inexplicable reasons. Since cardiovascular disease has been shown to be associated with cognitive impairment, detecting and, whenever possible, treating structural abnormalities of the heart before they become irreversible or spin out of control seem a sensible strategy to delay cognitive deterioration and potentially Alzheimer's disease. For this to happen, cardiologists need to take a more active role in managing the effects of the heart on cognitive function. This is especially true in the elderly patient being managed for cardiovascular disease, where consults with neurologists, neuroradiologists, gerontologists, neuropsychologists and other specialists who deal with dementia prevention can offer counsel.

The Heart as the Instigator of Cognitive Collapse

We have known since the Ebers papyrus in 1552 BC, and probably even before, that the brain and heart are intimately connected. The ancient Greeks and Aristotle in particular believed that the function of the brain was to "cool" the blood, while the heart was the source of memory. This belief was solidified by religious and scientific dogma for centuries. It took the most significant achievements in medicine in the sixteenth and seventeenth centuries by the Belgian anatomist Andreas Vesalius, the Spanish anatomist **Miguel Cervet,** and the English physician **William Harvey** to challenge that prevailing dogma and describe a more accurate account of the cerebral circulation as well as the heart's continuous pumping action inside a very precise circuit.

Fast-forwarding to the twentieth century, several researchers in the late 1970s became aware of an intriguing link between a sick heart and the start of cognitive deterioration that often led to stroke and the development of vascular dementia. This link came to be known as "**cardiogenic dementia**," and although it remained largely ignored for many years as an irrelevant observation, it eventually opened the door slightly to a fascinating field where cognitive impairment and dementia could be triggered by a bad heart [40].

Research in the early 1990s additionally began to suggest that cardiovascular disease could also signal the start of Alzheimer's disease, an assumption that, as will be seen below, has gained evidence-based support from more detailed studies published in the last decade. **The suggested link between cardiovascular deficits as a risk for Alzheimer's was crucial not only because it has now been generally accepted to be true but also because it implied that other extracardiac vascular risk factors might also play a crucial role in the initiation of this dementia by possibly sharing common pathological pathways or markers common in Alzheimer's.** Moreover, the therapeutic implications of controlling such vascular

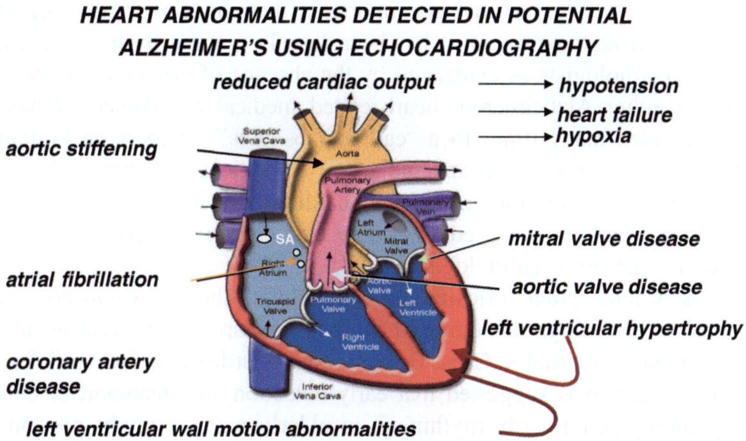

Fig. 12.8 Cardiovascular risk factors have been reported to promote progressive cognitive decline leading to dementia in the elderly population as detected by echocardiography. Reduced cardiac output (<3.4 L/min) can promote hypotension, heart failure, hypoxia, and increased vascular resistance. These risk factors are associated with cognitive decline (Fig. 12.1a–c) and ostensibly Alzheimer's. Aortic and mitral valve disease can promote blood flow regurgitation which contributes to low cardiac output. Similarly, valvular thickening can also impair normal cardiac output. Aortic stiffening is associated with aging, hypertension, and atherosclerosis and is believed to result in brain microvascular damage, leading to cerebral hypoperfusion and cognitive decline. Left ventricular wall motion abnormalities mainly result from myocardial ischemia. Atrial fibrillation is a risk factor for cardioembolic events, stroke, and for Alzheimer's; it is the most common arrhythmia in the elderly population. Left ventricular hypertrophy may be asymptomatic, mild, moderate, or severe and is a reported risk factor for cognitive decline in middle age but loses its predictive value in advanced age. Epidemiological and clinical evidence indicates that coronary artery disease, the leading cause of mortality in the USA, is a potential risk factor for Alzheimer's. Presence of two or more cardiovascular risk factors may significantly accelerate the onset of cognitive impairment

risk factors for Alzheimer's have become one of the most important inroads in the search to lower the rising prevalence of either Alzheimer or vascular dementia.

Cardiovascular disease comes in many forms, and its outcome is dependent on the patient's age, personal and family history, lifestyle, primary prevention, genetics, and pathological factors affecting structural and hemodynamic function (Fig. 12.8). Cardiovascular disease influences the heart vessels which in turn affect the structural tissues that make up the heart. These structural abnormalities can punish not only the coronary vessels but also any of the 4 valves that allow blood to flow in and out of the heart. In addition, structural pathology can damage the electrical system that controls the heart rate and rhythm, the size and function of the heart's 4 chambers, and the amount of blood pumped out with every beat.

Although the brain is the commander in chief of our body, the heart is the generator that keeps all mental and physical activities in good working order. It seems rather obvious that if something impedes the heart's action from accomplishing its duties, something, somewhere in the body is going to suffer and the most likely candidate is the brain because of its high metabolic demand for high energy nutrients.

Consequently, despite the obvious importance that connects some types of heart disease to cognitive impairment, it has been largely ignored by many researchers and by most cardiologists as evidenced by the absence of presentations discussing this topic in either Alzheimer or heart-related medical conferences. I have personally responded many times to a "call for abstracts" which is an invitation to present data or discuss a relevant subject at a medical conference, but all my abstract submissions were turned down by cardiology meeting organizers because my topic concerning cardiovascular disease and cognitive impairment was considered not of interest to cardiologists.

This lack of intellectual curiosity and comfort with the *status quo* goes back to the time when an anonymous editorial in 1977 appeared in the Lancet, the most prestigious medical journal in Great Britain, titled "Cardiogenic Dementia" [40]. In that brief article, it was suggested that early detection and treatment of abnormal cardiac pumping activity (dysrhythmias) in elderly patients might prevent intellectual deterioration in these senior patients. The dysrhythmias included atrial fibrillation and paroxysmal tachycardia, and the reason was that the elderly brain is highly sensitive to low oxygen delivery from improper cardiac pumping. The Lancet article went on to point out that senile dementia (which later became Alzheimer's disease) appeared to be aggravated or even caused by cardiac abnormalities. Finally, the Lancet article stressed that the relationship between heart disease and dementia seemed very important and worthy of further study by cardiologists [40]. The Lancet article was promptly ignored, and the study of cardiovascular disease and cognitive impairment languished in medical library shelves until recently, but not in cardiology journals. There appear to be scant few papers that have questioned whether cognitive decline generated by low cardiac output in the mildly symptomatic elderly can lead to Alzheimer or vascular dementia.

Most articles that appear in medical journals concerning the association between cardiovascular disease and cognitive impairment are published in neuroscience journals. It is not clear why they are not published in cardiac-related journals. The reason for this attitude is puzzling. Moreover, very few studies have examined the role of cardiac disease as a source of brain hypoperfusion despite intriguing evidence that the presumed mechanism points to reduced cardiac output from left ventricular dysfunction, a condition called heart failure.

Cardiology journal editors would do well to enter twenty-first-century thinking concerning the vital role played by the heart in cognitive function and dysfunction. This engaging action would expand the limited knowledge shown or ignored by cardiologists concerning the role of cardiovascular disease on cognitive function and would be welcomed by patients who would benefit from any information their cardiologist could provide to help them delay cognitive deterioration and consequential dementia. Currently, this information is not routinely discussed with elderly patients with impending or active cardiovascular disease in the USA. Nor are referrals to neurologists or Alzheimer specialists part of the general cardiology visit when cardiovascular risk factors encountered. This problem is discussed in detail in Chap. 16, under Heart-Brain Clinics.

Heart Failure

Steady, prolonged, and unrelenting damage to the brain can occur from low cardiac output. The process involves uninterrupted hemodynamic pump dysfunction and can significantly lower blood flow to brain cells, thereby diminishing energy substrate supply needed to conduct normal neurometabolism.

Heart failure, also called congestive heart failure, is the most common reason for hospitalization among older adults. In the USA, close to 6 million people are affected with about 50,000 new cases recorded every year. It is one of the fastest growing cardiovascular disorders due to the rising elderly population and is estimated to affect 20 % of people over age 80. Heart failure is a condition in which the heart cannot adequately pump enough blood to meet the body's needs. It is characterized by shortness of breath, fatigue, and buildup of tissue fluid especially in the legs, ankles, and feet. The most common cause of heart failure occurs from stenosis of the coronary arteries which supply oxygen to the heart. Heart failure is often associated with other comorbid vascular risk factors for Alzheimer's including ischemic heart disease, hypertension, and atrial fibrillation. Brain hypoperfusion is a common outcome of heart failure because the heart is unable to pump the blood and oxygen necessary to meet the brain's demand. As blood flow pumped out of the heart slows down, returning venous blood to the heart backs up, causing tissue swelling particularly in the lungs. Heart failure is the most common reason for hospitalization among older adults and has been repeatedly reported to worsen cognitive impairment and increase the risk of Alzheimer's.

Some medical reports have indicated that heart failure in elderly persons is associated with lowered cerebral blood flow in the posterior cingulate gyrus and the lateral temporoparietal cortex, regions that are linked to memory and visuospatial orientation [41]. This finding is of interest since memory and visuospatial dysfunction are two of the earliest signs in imminent Alzheimer's.

Other reports [42] have observed that heart failure patients with a dysfunctional left ventricle resulting in low cardiac output show an association with cognitive impairment. A normal cardiac ejection fraction which can be determined using noninvasive echocardiography ranges between 55 and 70 %, which is a percentage of how much blood is being pumped out of the left ventricle. In heart failure, an ejection fraction less than 40 % can confirm a diagnosis of heart failure, and if this volume falls below 35 %, life-threatening irregular heartbeats can follow. In some cases, heart failure can exist even when the ejection fraction is normal.

Heart failure has been shown to be an important precursor of impaired reasoning, memory, psychomotor speed, and executive functions. These qualities define us as humans, and without them, practical existence is improbable.

Although brain perfusion and cognitive function are reduced in heart failure, this syndrome can improve following implantation of a pacemaker in patients with low heart rate or from the use of selective cardiovascular agents. Clearly, aggressive treatment of heart failure to reverse brain hypoperfusion could have a significant impact in reducing the incidence of Alzheimer's in these patients. Heart failure is often associated with other comorbid vascular risk factors for Alzheimer's, including ischemic heart disease, hypertension, and atrial fibrillation, where heart rate is unusually fast and heart rhythm is abnormal causing dizzy spells, fatigue, shortness of breath, and often strokes. Brain hypoperfusion is a common outcome of heart failure. As blood flow pumped out of the heart slows down, returning venous blood to the heart backs up, causing tissue edema particularly in the lungs.

There is an urgent need to assess and treat patients with heart failure to prevent cognitive dysfunction since this condition can affect the patient's ability to make appropriate self-care decisions.

Coronary Artery Disease

In the USA, over 380,000 deaths occur annually from coronary artery disease, also called heart disease, making this malady the leading cause of death for either men or women. There is no single cause of coronary artery disease. The pathology usually involves the buildup of cholesterol-rich plaques called atheromas on the inside lining of the coronary arteries. The atheromas tend to thicken the arterial wall causing narrowing of the arterial space and reduced delivery of oxygen and nutrients to the cardiac tissue. The atheromas are made up of a chemical *bouillabaisse* that includes cholesterol, fatty compounds, inflammatory cells, calcium, and fibrin. When blood supply and oxygen supply to cardiac muscles are restricted due to narrowing or blockage of coronary arteries, anginal pain or death can result (Fig. 12.8). Atheromas can occur in coronary, cerebral, and peripheral arteries.

When stroke occurs due to a thrombus from a carotid artery, a combination of factors involving clotting mechanisms and hemodynamic changes within cerebral arteries is at play. This interaction explains the mechanism of ischemic stroke in patients with carotid atheroma which may be due to artery-to-artery embolism or low cerebral blood flow (Fig. 12.8).

Not surprisingly, vascular risk factors for Alzheimer's such as hypertension, diabetes, smoking, obesity, sleep apnea, lack of exercise, and high levels of cholesterol and lipids are also risk factors for heart disease. These comorbid conditions are a close-knit family devoted to ending life or causing devilish harm as advanced aging approaches. The person whose philosophy is "live and let live" and "save your advice for the hypochondriacs" is being very foolish if the above conditions are present or knocking at the door. Some individuals may have 2 or more of these risk factors and survive well into the century club, but more often than not, to disregard the presence or potential appearance of these coronary killers is to entertain a death wish or life with an unresponsive brain. A personal and family history, blood tests, electrocardiogram, echocardiogram, and stress test are relatively easy, quick, and noninvasive office procedures (blood tests require one small needle stick) that can provide the physician with excellent information on how to treat, prevent, and manage most heart conditions from threatening life or cognitive impairment.

Atheromas are the basis of atherosclerosis in coronary, peripheral, or cerebral blood vessels. When an atheromatous plaque erodes or suddenly ruptures, platelets clump up around it, inducing a thrombus or increased narrowing of the vessel, a condition that can result in myocardial infarction or ischemic stroke. This process involves clotting mechanisms and hemodynamic pathology in the vessels affected by thrombi. The use of platelet deaggregators which prevent blood cells from clumping in blood vessels, such as aspirin, clopidogrel, and cilostazol, or drugs called phosphodiesterase type 3 inhibitors that can widen vessel lumen to increase blood flow, are generally used to prevent stroke consequences.

The use of tissue plasminogen activator (tPA) can break a fibrin clot in the brain and restores blood flow in acute ischemic stroke that threatens permanent neurological deficits. But, for tPA to work, it must be given within four hours after the onset of stroke. New oral anticoagulants (NOACs) such as rivaroxaban (Xarelto) and apixaban (Eliquis) are factor Xa inhibitors in the coagulation cascade and can be used to prevent thrombi to the brain caused by atrial fibrillation and deep vein thrombosis. These agents appear superior to the classic anticoagulants warfarin (Coumadin) and vitamin K antagonists with less serious side effects such as hemorrhage. NOACs do not require regular blood monitoring as other anticoagulants do.

Ischemic stroke consequences can be neurologically and mentally disabling. For example, body paralysis opposite to the hemisphere where the stroke occurs, visual problems, speech and language difficulties, and memory loss can follow a stroke. Stroke and coronary artery disease are known to reduce brain blood flow and besides memory loss can also impair other facets of cognitive function. Stroke and coronary artery disease are considered to be important vascular risk factors to Alzheimer's.

Valvular Disease

Few studies have examined the effects of myocardial valve damage and its possible effect on cognitive function. Autopsy findings have reported significant aortic and mitral valve damage in Alzheimer patients when compared to a non-demented control group [43]. This association linking valvular damage and Alzheimer's is consistent with the presence of brain hypoperfusion at the preclinical stage of the unrolled Alzheimer pathology [43], an observation supported by other findings [44]. Atrial fibrillation can develop from mitral valve damage involving either mitral valve prolapse or valvular regurgitation (Fig. 12.8).

Mitral valve prolapse and abundant senile plaques characteristic of Alzheimer pathology were found at autopsy in the brains of elderly subjects who died from critical coronary artery disease but were not demented. The link between senile plaque formation and cardiac valvular damage implies that structural cardiac damage may influence the formation of neurodegenerative lesions even when no dementia has yet been expressed. This may occur because damage or deficits affecting the mitral and aortic valves can diminish cardiac output over extended periods of time before the development of heart failure symptoms. As the heart pumps less blood to organs including the brain, slow onset of cognitive deterioration in aging individuals becomes a real possibility.

It is tempting to speculate that left atrial dysfunction stemming from mitral valve damage can be a therapeutic target for cognitive decline that is preventable either by surgical or by pharmacological approaches to lessen the risk of Alzheimer's in such patients. More work should determine whether our speculation has any merit.

Atrial Fibrillation

Atrial fibrillation is the most common heart rhythm disorder classified as a cardiac arrhythmia. It is most common in the elderly where its prevalence is growing due to an increased life span (Fig. 12.8). Atrial fibrillation occurs when the electrical impulse that normally starts at the sinoatrial node (SA) instead starts somewhere else in the atria becoming irregular and much faster than normal. This abnormal rhythm cannot squeeze blood effectively into the left ventricle. The abnormal blood flow to the ventricle makes the blood more likely to clot, primarily in the left atrial appendage, a muscular pouch that joins the left atrium. An atrial clot can then be pumped out of the heart and lodge in a brain artery causing a stroke. Oral anticoagulants (NOACs) are used routinely to prevent these clots from forming in the left atrial appendage.

In the normal heart, the rate of ventricular contraction is the same as the rate of atrial contraction. In atrial fibrillation however, the rate of ventricular contraction is less than the rate of atrial contraction. This condition can lead to a decrease in cardiac output diminishing the amount of blood pumped into the body by the ventricles. The reason is because the upper chambers of the heart, the atria, are unable to fill the ventricles adequately with blood due to their rapid rate of contraction and their absence of normal contractions. The risk of atrial fibrillation increases with age and is more common in males.

Not surprisingly, studies have shown an association between atrial fibrillation and diminished cognitive function leading to Alzheimer's. Other studies have reported that people with atrial fibrillation develop cognitive decline at a faster rate when compared to others without atrial fibrillation [45].

Atrial fibrillation also appears to involve a significant conversion to dementia in non-demented people whether or not cognitive impairment was present. This outcome could be due to a stroke leading to vascular dementia. Many studies have shown that atrial fibrillation induces significant brain hypoperfusion which can compromise the aging cerebrovasculature in a way that can affect the delivery of essential nutrients to brain cells.

Although the true mechanism that associates atrial fibrillation with cognitive impairment remains unclear, our suspicion is that brain hypoperfusion may be triggered by the chronic arrhythmia associated with atrial fibrillation which has the added potential of inducing microembolic brain infarcts.

Left Ventricular Hypertrophy

Left ventricular hypertrophy refers to an enlargement of the muscle wall in the left ventricle (Fig. 12.8). This can occur as a response to chronic, uncontrolled hypertension or some other cardiac factor. There is evidence that left ventricular hypertrophy can also be influenced by genetic factors. However, the underlying gene patterns and genes remain obscure although it is suspected that multiple interacting genes with the environment may explain the genetic component of this cardiac anomaly.

The main problem with left ventricular hypertrophy is that the heart needs to pump harder in order to overcome the reduced elasticity of the thickened ventricular wall which tends to weaken the pumping force of the heart. When the problem relates to high blood pressure, treatment with antihypertensive therapy is indicated. Not treating this condition can lead to cognitive deterioration and Alzheimer's in the elderly as well as a higher mortality and morbidity from cardiac disease [46].

Left ventricular hypertrophy is a predictor of stroke and heart failure. When and if the condition worsens to a point where the heart pumping force is inadequate to maintain blood flow to the brain and other organs, a ventricular assist device may be needed. A left ventricular assist device is a mechanical pump used to support the function of the heart, so it maintains an adequate blood flow in people who have weakened hearts. In other circumstances, a heart transplant can also be considered for severe left ventricular hypertrophy.

However, good management of left ventricular hypertrophy is presently achievable with antihypertensive therapy and poses a good prognosis in those patients treated.

Hyperhomocysteinemia

Hyperhomocysteinemia is a medical condition characterized by increased levels of homocysteine in the blood. Homocysteine is a sulfur-containing amino acid that is not obtained from foods directly but through a complex series of biochemical steps involving methionine, another common amino acid.

Hyperhomocysteinemia is linked to a number of disorders associated with blood vessel inflammation, atherosclerosis, heart attack, and stroke. It is also a promoter of deep vein thrombosis, a condition that often leads to fatal pulmonary embolism when a clot from the leg is dislodged to the pulmonary arteries. Elevated serum levels of homocysteine are associated with cognitive loss, possibly through its effect of reducing blood flow resulting from narrowed and damaged arterial feeders to the brain. Low intake or deficiency of vitamins B12, B6 and folate appear to contribute to hyperhomocysteinemia, but replenishing these vitamin deficits does not lower the incidence of heart attacks or strokes when homocysteine serum levels are high [47]. However, one small but randomized, double-blind controlled trial from investigators at the University of Oxford reported that a high dose of homocysteine-lowering vitamins consisting of folic acid and vitamins B6 and B12 were able to slow down progression of mild cognitive impairment in an elderly cohort. The study was done over a two-year period and compared multivitamin-treated to a placebo-treated group [48]. The authors speculated that lowering homocysteine levels could be a preventive measure in delaying the onset of Alzheimer's.

Other studies have not shown that multivitamin supplements (B12 and B6) can improve cognition in older adults [47]. Lowering homocysteine as a treatment to prevent heart attacks remains controversial [47].

A pioneer in the discovery demonstrating that high level of homocysteine was an independent risk factor for cardiovascular disease was **Dr. Kilmer McCully** who in 1969, showed evidence that patients with homocysteinemia developed arterial and arteriolar lesions in various parts of the body including the brain [49]. McCully attributed these vascular changes to **damage of the endothelial cells which line small arteries** and are now known to be of pivotal importance in controlling cardiac and brain blood flow. McCully also observed in patients with hyperhomocysteinemia a thickening of the carotid arteries which we now refer to as intima–media thickening. This vascular pathology is presently seen in preclinical Alzheimer patients who develop carotid artery narrowing during advanced aging [50]. These early observations have been amply confirmed in many studies carried out in the last two decades [51].

McCully's observations have added considerable value to our understanding of the role played by the heart and the brain vasculature in promoting reduced blood flow to either organ and the eventual effect on cognitive dysfunction.

Endothelial cell damage caused by hypercysteinemia in brain vessels has not been well studied so far but could explain the pathology related to the narrowing of cerebral arteries that are prevalent in Alzheimer brains. If this conclusion is shown to be valid, therapy to reduce homocysteine in the blood could be a useful adjunct to help prevent narrowing of intracerebral arteries. This therapeutic target might also help the compromised cardiovascular circulation and extracerebral vessels supplying blood flow to the brain.

The role played by hyperhomocysteinemia in cognitive impairment and dementia is a research area receiving intense interest. This interest extends to other neurological disorders such as Parkinson's disease, multiple sclerosis, peripheral neuropathy, diabetic neuropathy, and epilepsy. Determining how these medical conditions relate to increased homocysteine abnormalities and cerebral blood flow deficiency will be an important contribution to our understanding of how to clinically manage them. Although links between hyperhomocysteinemia and cardiovascular disease have been reported, the role of this amino acid on cardiac output, needs to be addressed.

Nonetheless, we are aware of scant few papers that have questioned whether cognitive decline generated by low cardiac output in the mildly symptomatic elderly with hyperhomocysteinemia can lead to Alzheimer's or to vascular dementia. Moreover, very few studies in the Alzheimer literature have examined the role of cardiac disease in the presence of high serum levels of homocysteine to be a source of cerebral hypoperfusion despite evidence that the presumed mechanism leading to cerebral hypoperfusion secondary to reduced cardiac output is, ostensibly, left ventricular dysfunction associated with a reduced stroke volume [52]. Everything considered, the ultimate fallout from low cardiac output on the brain is that hemodynamic pump dysfunction in the older person can significantly lower blood flow to brain cells as a result of brain hypoperfusion, thereby diminishing energy substrate supply needed for normal brain cell metabolism. This may or may not involve hyperhomocysteinemia.

Hypertension

Hypertension is now recognized as the most important risk factor for Alzheimer's and vascular dementia. An important clue about essential hypertension in the US population is that it is uncommon in young adults but by the age of 65–75, its prevalence increases by 60–70 %. This statistic takes on added significance when it is realized that the proportion of people older than 65 will dramatically increase within the next decades.

Great strides have been made in the last decade to identify major risk factors for Alzheimer's disease (see Fig. 12.1a–c). A number of clinical studies indicate that hypertension appears to be a major risk factor for Alzheimer's and to vascular dementia and its control could prevent or significantly delay cognitive dysfunction from progressing to either dementia. This association is crucial because 24 % of the US adult population representing over 43 million people have hypertension. What is unclear are the cellular–molecular mechanisms that can lead to cognitive decline and frank dementia in hypertensive individuals. It is generally recognized that midlife hypertension often precedes Alzheimer's or vascular dementia, either following chronic brain hypoperfusion or following an ischemic stroke. There is a general agreement that hypertensive-related brain deficits are tied to cerebral blood flow and oxygen metabolism. An important but unresolved issue is whether *chronic* hypertension promotes cognitive impairment in the elderly *prior* to Alzheimer's by triggering endothelial cell vasoconstrictors or by manufacturing intracellular oxidative stress molecules that lower brain blood flow to critical levels through mitochondrial damage.

Molecular mechanisms affecting oxygen transport to cells may provide a clue. Hypoxia-inducible factor 1 (**HIF-1**) is a transcription factor that regulates adaptive responses to reduced oxygen in mammals. At the systemic level, hyperventilation can improve oxygen delivery to vital organs such as the brain, but when oxygen delivery is deficient due to chronic hypoxia, HIF-1 may be expressed as an adaptive mechanism in the hypoperfused brain by upregulating its downstream signal, vascular endothelial growth factor (**VEGF**), a powerful vasodilator and promoter of angiogenesis. HIF-1 is composed of HIF-1alpha (**HIF-1*a***) and HIF-1beta subunits that regulate genes to compensate for hypoxia, metabolic energy loss, and oxidative stress. HIF-1*a* is unique since its expression is tightly regulated by cellular oxygen concentration.

My personal belief is that HIF-1*a* acts as an early sensor of cerebral blood flow following hypertension and may promote improved brain perfusion via VEGF activation. This is a novel correlation not previously well studied. If HIF-1 expression is provoked by hypertension, a mechanistic link between blood pressure and VEGF activity could provide an alternate approach to treating hypertension prior to Alzheimer onset by increasing HIF-1*a* upregulation. Thus, my thinking is that sustained hypertension during aging is an important risk factor for Alzheimer's by evoking neurodegenerative trigger factors such as hypoxia, endothelial cell vasoconstrictors, and oxidative stress production. Consequently, the resulting cognitive impairment linked to hypertension may worsen in the presence of brain hypoperfusion. However, this pathological process could be partially reversed by promoting HIF-1*a–VEGF* expression using activators such as dimethyloxalylglycine (DOG) or activating transcription factor 4 (ATF4) that can lead to angiogenesis.

It has been estimated that middle-aged and older adults after age 50 have a probability of over 90 % to become hypertensive during their lifetime. These clues are especially relevant when one considers that regional cerebral blood flow shows brain tissue shrinkage and hypoperfusion in normal elderly individuals over age 60. Typically, a 20 % drop in cerebral blood flow occurs from age 20–65. Hypertension and advancing age are two big hits on maintaining adequate blood flow to the brain. The term "vascular aging" is relevant to hypertension since blood vessel aging is at least in part, responsible for high blood pressure seen in advanced aging. This process of aging blood vessels is also involved in atherosclerosis, cardiovascular disease, stroke, and arterial stiffening.

Arterial stiffening is a biologic marker of atherosclerosis and is characteristically associated with hypertension and other vascular risk factors involved in cognitive impairment. If untreated, it will result in reduced arterial distensibility, thus adding a third hit to cerebral blood flow decline. In view of this collective evidence, it is not surprising that the Baltimore Longitudinal Study of Aging led by **Dr. Shari Waldstein** and her colleagues [53] determined that hypertensive patients show poorer cognitive ability than normotensives on tests of memory, attention, and abstract reasoning. This conclusion has been verified in a majority of subsequent studies. The SYST-EUR 2 randomized controlled trial reported that antihypertensive therapy in a 4-year follow-up study significantly prevented cognitive decline prior to Alzheimer onset as compared to controls [54]. Findings from the HOPE [55], PROGRESS [56], and SCOPE [57] trials confirmed that antihypertensive intervention prevented cognitive decline associated with stroke or recurrent stroke in a significant percentage of patients.

It is also worthy to note that antihypertensive therapy given at middle age will be more successful than administering it during late life.

High blood pressure is known to reduce cerebral blood flow, but the mechanism involved in this action is unclear. Because cardiac output remains normal in high blood pressure, two possible causes for the reduction of brain blood flow may occur in hypertensive individuals. First, hypertension can increase systemic vascular resistance and slow down normal blood flow. Second, cerebral blood flow may fall when blood pressure is high owing to direct damage to brain endothelial cells that produce the vasodilator nitric oxide. Release of vascular nitric oxide from endothelial cells to produce vasodilation is a well-established reaction. Vascular nitric oxide has been shown to diffuse toward the lumen of blood vessels in humans where it helps maintain blood fluidity and thereby reduces blood viscosity and resistance, thus improving blood flow. Vascular nitric oxide production can be compromised when brain hypoperfusion occurs.

Therefore, there is an implied possibility that therapy aimed at controlling high blood pressure using potassium-sparing diuretics or increasing cerebrovascular nitric oxide or protecting endothelial cells in the brain pharmacologically by limiting low mean shear stress and oscillatory flow could prevent or delay the onset of Alzheimer's by counteracting chronic brain hypoperfusion in the hypertensive population. In fact, increasing vascular nitric oxide in the brain of hypertensive people should provide another benefit. Vascular nitric oxide activation, for example, has been shown to lower Abeta production from being enzymatically cleaved by BACE-1. BACE-1 is the initial proteolytic enzyme responsible for Abeta synthesis. At the same time, vascular nitric oxide increases the synthesis of BACE-2, the enzyme that cleaves amyloid precursor protein (APP) so that Abeta production cannot take place [58]. Although there is no guarantee that increasing vascular nitric oxide synthesis in the brain will benefit hypertension, there are a number of ways this can be examined and the reader is referred to a review on this topic for details [59].

These are theoretical considerations that need to be explored experimentally. Hypertension is well suited for prevention by an armamentarium of pharmacological and behavioral interventions, including antihypertensive therapy that best fits the patient and by lifestyle changes that involve physical exercise, nutritional foods that lower cholesterol level and salt intake, cessation of smoking, control of obesity, and techniques to lower physical and mental stress.

On a more technical level, hypertension can cause a seesaw disbalance between vasoconstriction and vasodilation of arterioles in the brain. This disbalance is due to endothelial cell dysfunction that may affect cerebral microvessels. Endothelial cells are primarily damaged by the production of oxygen free radicals that can compromise vascular nitric oxide biosynthesis and impair its vasodilator function and its protective effects on the arterial wall. Endothelial cell damage is commonly associated with chronic hypertension and also with an increased production of the vasoconstrictors thromboxane A_2, prostaglandin H_2, 20-HETE, and superoxide anion in cerebral resistance arteries. These molecules are often precursors and co-conspirators of oxidative stress.

Oxidative stress has been defined as an imbalance between the excessive production of toxic free radicals and the inability by the body to counteract the harmful effects of these molecules. In addition, oxidative stress can negatively impact many tissue types, including vessels in the brain and heart that can lead to cerebro- and cardiovascular diseases. Evidence is available that persisting oxidative stress from hypertension results in limiting nitric oxide synthesis from nitric oxide production, a feature that would produce the toxic tissue molecule called superoxide anion. As the "spark plug igniter" of many disorders, superoxide anion has been shown many times to provoke the production of molecules associated with Alzheimer's, stroke, and heart disease.

Hypotension

In clinical practice, very little importance is attached to a patient with hypotension. Medical dogma has classically considered that low blood pressure should not be treated because cerebral autoregulation prevents reduced brain blood flow from becoming harmful. This doctrine is not accurate. Many recent studies have shown that elderly hypotensive individuals pose a significant risk factor for developing cognitive failure and Alzheimer's [7]. Not only can brain perfusion become impaired in hypotensives, but it can also induce lack of drive, fatigue, headaches, and faintness. These mental and organic changes have been confirmed using psychometric and physical examinations.

It would appear that diastolic pressures of less than 65 mm Hg may be associated with a greater chance of developing Alzheimer's [7], but this finding needs to be further confirmed in clinical trials.

Although there is substantial evidence that hypertension is linked to cognitive decline (see section on **Hypertension** above) in the elderly and should be treated, an important caveat is that the use of aggressive antihypertensive medication can lower blood pressure to a level where the elderly individual is just as prone to cognitive dysfunction as if hypertension were present. This paradox suggests that a thin line separates too high and too low blood pressures from cognitive disability.

Non-antihypertensive pharmacotherapy is an alternative approach that may be successful in lowering hypertension. Activities such as daily walking, mild–moderate exercise, coupled with reduction of salt intake, stop smoking, lowering cholesterol by avoiding as much as possible lipid-laden foods have been shown in numerous studies to lower the incidence of Alzheimer's. When these measures are not realistic or complied with, patient home monitoring of blood pressure and reporting results periodically to a healthcare provider may keep the blood pressure at an optimal range.

The World Health Organization defines low blood pressure as a systolic reading below 110 mm Hg in males and below 100 mm Hg in females, regardless of diastolic blood pressure. While the reason for this guideline cannot be explained by common sense, it is known that such hypotensive people can suffer from faintness, fatigue, reduced drive, and headaches. It has also been reported that hypotensives are at higher risk of cognitive impairment and Alzheimer's than normotensive people [7].

Low diastolic blood pressure is associated with an increased risk of Alzheimer's in the elderly population, particularly among users of antihypertensive drugs [7]. While the reason for this finding is not totally apparent, undetected chronic brain hypoperfusion from high and low blood pressures could be a key pathogenic link to Alzheimer's dementia.

White-Coat Hypertension

A common office procedure that has received almost no clinical attention in relation to Alzheimer's development may provide a clue to blood pressure ambiguities. That procedure is white-coat hypertension. This problem occurs when a transient and usually mild increase in blood pressure is observed in certain individuals when attending a doctors' office. White-coat hypertension is not generally linked to target organ damage or prognosis from true normotensives. Although there is a controversy about whether white-coat hypertension should be treated or not, many doctors assume mild blood pressure elevation needs to be treated, especially in the elderly. As a consequence, these individuals are given antihypertensives for the rest of their lives which can lower their blood pressure **subnormally**. It must be remembered that in the aging individual, diastolic hypotension can introduce not only chronic brain hypoperfusion, but also the prospect of Alzheimer's. Because the results of many studies are consistent with the view that white-coat hypertension is not likely to be improved by drug treatment, a white-coat hypertension reaction needs to be evaluated with ambulatory pressure. If ambulatory pressure is normal in the mild-hypertensive white-coat hypertension may be suspected and antihypertensive therapy would be contraindicated. Several studies have shown that the main effect of antihypertensive drugs is to lower the clinic pressure, without having a significant effect on the ambulatory pressure, which by definition is normal to begin with.

The review of cardiac anomalies in older people, as reviewed above, pushes me to consider cardiovascular disease and its pathological ramifications, as the "smoking gun" in the physiochemical cascade that leads to cognitive loss and Alzheimer's or vascular dementia. What the heart appears to represent in the clinical picture preceding dementia is the generator pump that keeps the body organs and tissue in survival mode. When this pump begins to fail, even mildly, dementia is one of many consequences that may be averted as long as the failing pump can be therapeutically managed, repaired, or replaced.

We should now look for ways to aggressively reduce or reverse low cardiac output, abnormal blood pressure, coronary artery disease, arrhythmias, valvular disease, and markers of heart disease that can predict cardiovascular events, especially in the elderly [60]. Measures of efficacy for applied interventions need to be coupled to detected cardiovascular abnormalities using simple clinic office tools, such as noninvasive echocardiography that can be used cost-effectively to prognose or delay a host of heart-related anomalies (Fig. 12.8).

As reviewed in this section, epidemiological findings indicate that a broad spectrum of cardiovascular risk factors, including heart failure, thrombotic events, hypertension, hypotension, homocysteinemia, hypercholesterolemia, high C-reactive protein, coronary artery disease, valvular disease, heart failure, and atrial fibrillation, are more common in the elderly.

These conditions are reported to contribute to cognitive dysfunction and decline affecting performance in executive functions, attention, learning, psychomotor speed, verbal fluency, mental alertness, and memory. Consequently, the neuropathological link between the cardiac abnormalities listed above and their satellite offshoots, such as amyloid angiopathy and presence of the ApoE4 genotype, appears to contribute to a pathological vascular complex that appears to target Alzheimer's dementia via specific cardiopathic pathways.

I and others suspect that the risk for Alzheimer's from heart disease could originate primarily from atherosclerotic plaques within coronary vessels that damage cardiac endothelial cells lining the inside wall of heart vessels. Cardiac endothelial cells are capable of releasing chemical substances that play a pivotal role in regulating coronary artery blood flow. When this happens, the heart's pumping ability can be slowed to a point where poor blood flow to the brain can compromise neurons and neural tissue. This thinking is supported by the presence of high levels of cholesterol, low-density lipoprotein, and triglycerides found in the blood of "at-risk" Alzheimer patients. These lipids can accumulate inside coronary blood vessels and reduce blood flow to the brain, sometimes for decades without being detected.

A number of studies are now in progress testing whether cholesterol homeostasis and lipoprotein disturbances using cholesterol-lowering statins can alter Alzheimer pathology. However, the results of these studies are inconclusive so far and to a degree, very controversial with regard to the potential neuroprotective effects claimed by using statins. Another approach that may protect against Alzheimer's is to pharmacologically increase brain or cardiac endothelial nitric oxide, a powerful vasodilator with antithrombotic, anti-ischemic, and ant atherosclerotic activities released by endothelial cells lining the arterial walls. This can be done pharmacologically with oral L-arginine, a precursor of nitric oxide and by ascorbic acid.

Endothelial production of nitric oxide is vital for controlling arterial diameter (also called vascular tone), as well as arterial pressure, platelet clumping on vessel wall, and growth of the spindle-shaped cells called smooth muscle cells lining the outside wall of arteries, which also contribute to vessel tone. It has been reported that plant foods, such as beets, arugula, kale, and spinach, are loaded with nitrates and nitrites capable of stimulating the production of nitric oxide in the body. Added

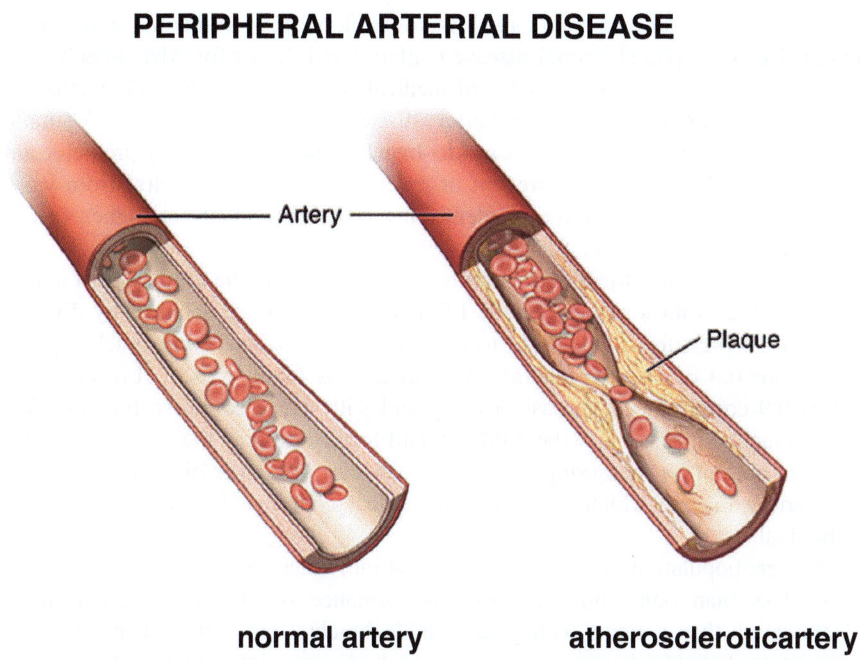

Fig. 12.9 Formation of atherosclerotic plaque in peripheral arterial disease can slow down blood flow to the brain and may be the primary reason why this disorder is a risk factor for cognitive decline and Alzheimer's. Atherosclerotic plaques can form in any artery to block or reduce blood flow. When they occur in cardiac or brain vessels, the rate of heart attacks and strokes is increased

to this list of nitric oxide boosters are flavonols and polyphenols contained in red wine, grapes, cocoa, and green tea [61]. This list of foods and others could very well be backed up in the future by rigid scientific studies proving their merit in helping to wall off some of the dangers that lurk in our environment, foods, air, and life stresses. For now, they can be regarded as sprinkles of common sense to be served and eaten with other nutritious and healthy foods.

Peripheral Arterial Disease

Peripheral vascular disease or peripheral arterial disease is a slow-forming, progressive narrowing or blockage of any vessel or lymph vessel outside the heart. The term peripheral arterial disease or peripheral artery disease is used when only arterial vessels outside the heart are affected. In the USA, peripheral arterial disease affects about 5 % of people over age 50 and about 20 % for those older than 70.

Peripheral arterial disease is a marker for coronary artery disease and carotid artery stenosis because atherosclerosis is often the initiator of narrowing or occlusion of the arterial vessel (Fig. 12.1a). Since narrowing or occlusion of vessels that supply the heart and brain can slow down blood flow, numerous studies have reported that peripheral arterial disease confers a risk factor for Alzheimer's.

There are a considerable number of medical conditions and lifestyle habits that favor the development of peripheral artery disease as shown in Fig. 12.9. A simple, quick, and noninvasive test called **ankle–brachial index** is an early predictor and diagnostic tool that can help prevent or reverse the affected vessel(s) from worsening and is a useful marker in identifying older individuals at higher risk of cognitive impairment (Fig. 12.5).

This test compares the blood pressure from the ankle to the arm. Normal ankle–brachial index values range between 1.0 and 1.4. Values are obtained by dividing the systolic pressure in the ankle to that of the arm. When ankle–brachial index values are 0.9 or less, peripheral artery disease is likely present. Lower values below 0.9 correlate with disease severity and with worse cognitive function. The ankle–brachial index is also the gold standard for determining the degree of arterial stiffness present by measuring pulse wave velocity, a marker of subclinical coronary artery disease which can eventually affect cognitive function in the aging individual.

A large, population-based study found that individuals having an ankle–brachial index less than 0.90 showed poorer performance on the Mini-Mental Status Examination than patients with greater ankle–brachial index [62]. These findings argue strongly for further screening patients who have peripheral artery disease with neuropsychological testing to assess cognitive abilities. The ankle–brachial index is a valuable tool when combined with other tests to find out whether peripheral arterial disease is present or predictable, but it is not perfect. Gender and ethnic differences have been reported as well as size of the patient (taller people have a higher index), and the examiner must take into account the age of the patient since older individuals tend to have greater arterial stiffening. Since the ankle–brachial index is the ratio of two pressures, it is not affected by blood pressure changes.

Of the ten conditions listed as risk factors for peripheral arterial disease (Fig. 12.1c), all are amenable for intervention to delay or prevent vascular disease.

Erectile Dysfunction

Erectile dysfunction is the term used to describe the persistent or recurrent inability by a man to maintain an erection sufficient to satisfy sexual activity. Erectile dysfunction is a common disorder in the male population affecting about 150 million people around the world with 30 million in the USA. The prevalence and severity of erectile dysfunction increase with age and may double in number by the year 2025. Most of the risk factors for erectile dysfunction are listed in Fig. 12.1c because peripheral artery disease and atherosclerosis are generally the causal basis. Other causes of erectile dysfunction include medications to treat depression and hypertension as well as stress and psychological issues. Less common causes of erectile dysfunction are penile injury and lower testosterone levels.

For this review, the relationship between erectile dysfunction and vascular disease is appropriate from the interventional point of view since correcting erectile dysfunction could be a key to preventing heart disease and stroke in the future. That action might also prove useful in preventing cognitive impairment. Correcting erectile dysfunction means addressing some of the risk factors for peripheral arterial disease listed in Fig. 12.1c. For example, cessation of cigarette smoking, a powerful vasoconstrictor, would be very helpful as would the lowering of alcohol intake and its neurodepressant effects.

Direct oral treatments for erectile dysfunction include drugs such as sildenafil, vardenafil, and tadalafil (Viagra, Levitra, and Cialis) that can increase the release of vascular nitric oxide in the penile tissue resulting in vasodilation and erection. These erectile dysfunction drugs, which are called phosphodiesterase-5 inhibitors, have also been looked at as possible treatments for declining cognition in Alzheimer animal models with reported success. It was found that sildenafil (Viagra) given to aged mice that produce excess Abeta in their brains was able to reverse the memory deficits induced in these animals, but this memory improvement was not related to a reduction of Abeta [63]. These findings suggest that sildenafil or other phosphodiesterase-5 inhibitors could have a beneficial effect on memory and cognitive decline in humans by boosting peripheral artery vascular nitric oxide as it does in the penis. Much more research is needed to confirm this possibility.

Sleep Apnea

A more recently recognized risk factor for cognitive decline in the elderly is obstructive sleep apnea syndrome. Sleep apnea is considered a risk for cognitive decline due to lowered oxygen (hypoxemia) during sleep and wakefulness. The degree of hypoxemia during sleep has been shown to be associated with impaired performance of concentration, attention, complex problem solving, and short-term recall when testing was performed on sleep apnea individuals.

Cognitive impairment appears commonly in patients with this sleep disturbance. Neuroimaging studies have been performed on sleep apnea patients to try and find out the brain regions that may be affected by sleep apnea. A common finding in several of these neuroimaging studies has shown a decrease in hippocampal volume, the region of the brain responsible for memory and learning. Chronic hypoxemia is the main deficit of sleep apnea. By not carrying sufficient oxygen in the blood, the major energy fuel for cells, ATP, becomes deficient, and in the brain, it can cause neurotransmission and synaptic deactivation that may lead to cognitive decline and dementia. Many of the peripheral artery risk factors for Alzheimer's (Fig. 12.1c) can promote sleep apnea, including cigarette smoking, diabetes type 2, alcoholism, and obesity. Controlling these conditions and the use of CPAP (continuous positive airway pressure) can reverse or adequately control sleep apnea and its consequences.

There is an unusual response during sleep to blood pressure changes that is considered normal. During sleep, blood pressure typically dips more than 10 % of daytime blood pressure. This is a normal compensatory activity that reflects a healthy, reactive vasculature is at work. People whose blood pressure does not dip more than 10 % from daytime baseline during sleep may show a higher risk for stroke or cardiovascular disease. Possible cognitive decline in elderly individuals due to an impaired vascular reactivity has also been proposed but remains unconfirmed.

Diabetes Type 2

Diabetes type 2 or adult-onset diabetes is a chronic metabolic disorder where abnormally high glucose levels appear in the blood as a result of relative insufficient insulin available. Its incidence in older people is similar to that of sleep apnea. This chronic disease process affects the function of many organs including the kidneys, heart, and nervous system.

Diabetes type 2 has been known for some time to be associated with executive function and memory deficits [64]. How it accelerates cognitive decline and the risk of dementia remains controversial. One point of agreement among investigators is that diabetes type 2 causes microvascular complications that can give rise to retinal and cerebral changes that can negatively affect mental flexibility, verbal fluency, and executive function. These preclinical conditions are the precursors of Alzheimer dementia. This type of diabetes can also damage larger vessels including coronary, peripheral, and cerebral arteries by promoting atherosclerosis, ischemia, and occlusion.

Glycemic control of diabetes type 2 is the preferred method of management. Glucose monitoring before and after meals is essential, and many drugs are available for this purpose. When management of glycemic levels is successful, diabetes-induced cognitive deterioration and microvascular or macrovascular abnormalities can be happily prevented.

Hemorheologic and Hemodynamic Abnormalities

The science of how blood flows and how it behaves following the laws of physics comes under the realm of hemorheology and hemodynamics. The laws of physics are the forces that determine whether blood flow will be adequate to organs or tissues or deficient.

Hemorheology and its close cousin hemodynamics are probably the primary elemental vascular disturbances that will slowly but relentlessly initiate the pathological cascade associated with dysfunctional cognition and later dementia. This sweeping statement can be made with some degree of confidence after 20 years of studying how blood flow to the brain behaves under normal and abnormal conditions in aged animals and humans [14]. Aside from my own observations, a considerable amount of data within the last two decades lend support to the role of insufficient blood flow to the brain and its consequences in older people. This crucial aspect of how Alzheimer's is first triggered in the elderly is discussed in Chap. 14.

For here and now, it can be said that blood flow is the most crucial constituent of life; no blood flow, no life.

Alcoholism

Anyone who has drunk too much alcohol at a single occasion in a party or a bar knows the effect it has on the brain especially in dulling the senses that result in lack of coordination, memory lapses, and confusion. It is not hard to realize that chronic and excessive alcohol intake will worsen the effects observed in an episode of alcohol intoxication on the brain. In older individuals, chronic alcoholism often leads to serious organic deficits which can be measured by partial loss of problem-solving ability, questionable judgment, lowered abstract thinking, poor psychomotor performance, and memory impairment. Excessive alcohol intake also called alcohol abuse impedes the digestive tract from absorbing many vital nutrients, including thiamine (vitamin B1). Thiamine deficiency can contribute to Wernicke–Korsakoff syndrome, a neurological disorder associated with severe amnesia and cognitive loss (see Chap. 1, **Wernicke–Korsakoff Syndrome**). The best treatment for alcohol abuse and reversing cognitive dysfunction is abstinence. Rehabilitation with counseling and an assortment of medicines are available to help the non-compliant patient remain sober and reduce the anxiety and poor sleep associated with alcohol withdrawal.

Menopause

Women have been shown to have a lower risk for peripheral vascular disease than men except when they reach and progress through menopause. Estrogenic deficiency may be responsible for this reduced protection, and the use of hormone replacement therapy to compensate has become controversial due to the increased risk of endometrial dysplasia and thrombogenic complications. However, some studies indicate that hormone replacement therapy in healthy postmenopausal women reduces the risk of peripheral arterial disease [65], while other studies challenge this conclusion [66]. Although there is no clear evidence so far for discontinuing hormone replacement therapy to prevent peripheral vascular disorders, the choice of whether to treat or not depends on ongoing research to further validate the reported benefits of hormone therapy.

Smoking

Compared to non-smokers, smoking more than 20 cigarettes per day for several decades has been linked to cognitive decline in midlife [67]. Cognitive dysfunction derived from heavy smoking is one of the most modifiable conditions of all the risk factors to dementia whether smoking is stopped at midlife. One study reported that ex-smokers who had ceased smoking for 10 years at midlife showed no adverse effects on cognitive decline and were no worse off than matched non-smokers [68].

Two main ingredients in cigarettes are especially toxic to the vascular system in humans, nicotine and carbon monoxide. Nicotine is an arterial vasoconstrictor and raises blood pressure speeding up the heart rate about 20 beats per minute with every cigarette. Carbon monoxide induces hypoxemia with every cigarette puff, thus increasing cardiac output. This outcome by nicotine and carbon monoxide and other toxins in cigarettes is a major risk factor for cardiovascular and peripheral arterial disease which will increase the chances of heart attacks, stroke, and Alzheimer's.

Hyperlipidemia

Lipidemia refers to fats in the blood. When normal lipid levels are present in the blood, they do important tasks for the body. When lipids link up with proteins, they form lipoproteins that provide energy for the body. Lipoproteins in the blood, also called cholesterol, are classified into three types: (1) high density (HDL), (2) low density (LDL), and (3) very low density (VLDL). High-fat foods, obesity, cigarette smoking, stress, alcoholism, diabetes type 2, and kidney and liver diseases will increase the "bad" LDL cholesterol levels that can generate atherosclerotic plaques in arteries and slow or block blood flow to organ tissue. To lower the effects of high LDL levels, reducing the intake of fatty foods, exercising, and avoiding, the risks factors for peripheral arterial disease (Fig. 12.1c) will have a positive effect on raising the "good" cholesterol HDL while lowering the "bad."

Healthy nutrition is one of the most important modifiers of hyperlipidemia for two reasons. First, it counteracts fat intake from non-nutritious diet. Second, it provides nutrients that can fight excessive fat intake and rising LDL.

Dr. M. Cristina Polidori at the University Hospital of Cologne and her colleagues have been at the forefront of this research. **Dr. Polidori has reported that Western diets generally favor saturated fat intake which can increase Alzheimer risk** [69]. Part of the remedy is to replace fish for meat and consume high amounts of fruits, vegetables, fresh dairy products, and plant-based fats and nuts [69]. Other food alternatives include replacing the saturated fat contained in butter or margarine with olive oil and salt with herbs and spices to flavor food. Drinking red wine in moderation is also recommended. This so-called **Mediterranean diet** has been shown to reduce heart disease and decreased risk of cognitive decline, including Alzheimer's [70]. More information on the Mediterranean diet can be found in a number of books on the marketplace, each with a small variation on the diet itself.

Obesity

Nearly two-thirds of adult Americans and one-third of children are obese or overweight. This is a jaw-dropping statistic. It has been estimated that 75 % of hypertensive cases are linked to obesity and 60 % to diabetes type 2, two major risk factors for Alzheimer's. Obesity is not just about Western cosmetics of looking good. It means having excessive body fat that increases the risk of many disease states and mental conditions, including cardiovascular, cerebrovascular, and cognitive disorders.

Major contributors of obesity are well known and include lack of physical exercise and bad food choices including high-calorie snacks, such as soft drinks, chips, chocolates, and oversize meals that generally include a good portion of *trans* fats. Less common causes but equally important contributors of obesity include medications, pregnancy, and genetic predisposition.

Obesity and overweight can be easily estimated by using the body mass index (BMI) which is computed by dividing the weight of the person by the height. A BMI of 25 or less is considered normal weight, while a BMI over 25–19 is considered overweight and a BMI over 30 is considered to be obesity, the higher the number indicates a higher degree of obesity.

Managing most cases of obesity due to poor nutrition is best done by dieting and physical exercise. Other methods may also be successful, but to keep the healthy weight from regressing to the obese state again, lifestyle changes need to be examined and followed. This is a matter of discipline and wanting to live longer. If the discipline to keep the excess weight off is successful, it is certain to have a positive effect on substantially reducing the chance of disorders that can lead to dementia through a multitude of predementia pathways. Disease states such as diabetes type 2 and hyperlipidemia are generally treated when detected and can have a salutary effect on controlling obesity.

The reader can now appreciate that the list of peripheral arterial disease conditions listed in Fig. 12.1c **is mostly interconnected in promoting cardiovascular and cerebrovascular damage**. As these risk factors appear during middle age and advanced aging, they can result in a direct pathway to cognitive difficulties, cognitive impairment and cognitive meltdown.

International Studies Focusing on Alzheimer's

The Finnish Geriatric Intervention Study (FINGER) is one of the few organizations in the world that investigates a multitude of therapies derived from many different medical disciplines. The rationale is to study an elderly population in order to search ways that may prevent or slow down cognitive impairment. The FINGER plan is to target several specific vascular risk factors in cognitively normal individuals in particular, for cardiovascular disease. The plan also aims to identify at an early stage elderly people at increased risk of cognitive decline and evaluate interventional strategies that may keep them remain cognitively functional.

Some of the diagnostic tools used in the FINGER study will be discussed in detail in Chap. 15. These tools include neuroimaging, neuropsychological testing, blood chemistry, echocardiography, ultrasound examination of the carotid artery, and pulse wave velocity measurement.

Besides FINGER, other international organizations are presently being organized to develop randomized controlled trials (RCTs) to prevent Alzheimer's. The European Dementia Prevention Initiative (EDPI), which includes FINGER, also has established two other multidomain preventive organizations including the Prevention of Dementia by Intensive Vascular Care (PreDIVA) and the Multidomain Alzheimer Preventive Trial (MAPT). The most important factor linking these organizations is to find preventive strategies in persons at risk of cognitive deterioration that can be slowed down with specific interventions. The factors under RCT scrutiny from all three RCTs are lifestyle changes related to Alzheimer's with an emphasis on physical activity, proper nutrition, cognitive training, and vascular risk factors.

The two-year interventional FINGER study was begun in 2014 and is still too early to evaluate any findings. It is made up of 1260 individuals randomized into two groups: lifestyle interventional changes and no intervention. The two groups share a similar average age of 69 and no significant differences in sociodemographic, vascular, or lifestyle characteristics [71]. These two groups did share vascular risk factors, namely systolic hypertension, high serum cholesterol levels, and similar Mini-Mental State Examination (MMSE). The FINGER study should provide useful information into the validity of introducing lifestyle changes in order to modify the potential outcome of cognitive decline and dementia.

PreDIVA is a Dutch study targeting 3500 non-demented elderly individuals aged 70–78 in a 6-year study and a 6-year follow-up. **Since the majority of elderly people have vascular risk factors that precede cognitive decline**, it is difficult to separate an intervention group from a control group given that both groups may be receiving vascular related therapy, for example, hypertension and high serum cholesterol. For this reason, PreDiva places importance not only on vascular risk factors but also on cognitive training and physical activity with an anticipated outcome of maintaining better cognition and a greater functional state in the intervention group.

The MAPT study includes 13 European centers enrolling 1680 elderly persons living independently and aged 70 or more. Inclusion into the study is subjective memory complaint which has not been diagnosed as mild cognitive impairment. The intervention period is for 3 years and a 5-year follow-up. The multidomain intervention emphasizes cognitive training, physical activity, and national guidance. One group in the intervention study will receive daily supplements of omega-3 fatty acids, which are reputed to lower high serum triglycerides and improve joint pain and muscular stiffness. The main goal of the study is to increase or sustain cognitive ability.

Preventive Alzheimer's Clinics

Not part of EDPI but just as important are individual centers for the prevention of Alzheimer's located primarily in Europe. The VAS-COG Clinic in Florence, Italy, was established as an outpatient clinic to assess patients for cognitive impairment secondary to cerebrovascular disease [72]. Patients are referred to the clinic by local physicians who may see patients with cognitive impairment and psychiatric issues derived from stroke or other cerebrovascular abnormalities. Patients who are evaluated at the clinic undergo a general and neurological examination, neuropsychological testing, and neuroimaging to detect cerebrovascular pathology [72]. Although no set of therapeutic standards presently exist for people at risk of dementia, the VAS-COG Clinic is a first step in the establishment of centers specifically dedicated to detect, evaluate, and formulate a plan of therapy or management for people at risk or already on the path to cognitive failure. As the science progresses, such centers become elemental in slowing down cognitive decline and giving patients a "grace period" of continued functionality not presently available anywhere else.

Another preventive Alzheimer's clinic which may be in the process of becoming a major center for dementia prevention is located at Leiden University Radiology Department, the Netherlands. This clinic focuses on potentially reversible vascular risk factors that induce the hemodynamic changes which can lead to cognitive impairment [73]. The foundation of this clinic is to create an interdisciplinary network around the country that engages in translational research, otherwise known as "bench-to-bedside" medicine.

The clinic's strategy is grounded on evidence-based medicine indicating that by improving the hemodynamics of poor blood flow to the brain, the resulting improvement of cerebral blood flow will parallel cognitive improvement. For example, it is widely accepted that stenosis of the carotid artery unilaterally or bilaterally can significantly lower brain blood flow in an elderly individual who will likely suffer from cognitive deficits. Surgical revascularization by endarterectomy may correct the stenosis providing improved brain blood flow and reversing the cognitive dysfunction. If the artery is blocked on the brain's surface, extracranial–intracranial bypass (usually by connecting the superficial temporal artery to the middle cerebral artery) can be performed to restore lowered blood flow and ostensibly diminished cognitive function. Although controversy surrounds extracranial–intracranial bypass surgery, the procedure may be helpful in a very selective group of patients.

Another example is heart failure. Heart failure is a condition that can severely lower cardiac output and brain blood flow. Heart failure is also a target at the Leiden clinic using pharmacological intervention such as the administration of angiotensin-converting enzyme inhibitors. Cardiac factors that influence cerebral blood flow, including blood pressure, aortic stiffness, cardiac output, and cerebrovascular autoregulation, can also be spotlighted in the dementia preventive process.

The Alzheimer Center Reina Sofia Foundation in Madrid, Spain, under the direction of **Dr. Jesus Avila**, is a research center that houses Alzheimer patients. These patients can be closely studied within the center's facilities offering valuable information on their disease progress and ways to maximize management. The foundation's research has shown that lower cardiac hemodynamic performance is significantly associated with structural brain damage that parallels diminished executive function and other cognitive brain abnormalities [74]. A large brain bank facility is the source of multiple neuropathological research studies.

The foundation also serves as a hub to connect other Spanish research centers, laboratories, and universities into a national and international network that fine-tunes medical and public attention, awareness, and resources toward limiting a disease of considerable consequences to society.

Is the System Rigged?

After reviewing the vast evidence linking cerebrovascular, cardiovascular, and peripheral artery disease with cognitive impairment, one is struck by the fact that **despite the obvious importance that connects these three systems to cognitive decline, it has been largely ignored by many researchers in the field and by most neurologists and cardiologists**. This conclusion is supported by the surprising absence of presentations discussing this topic in either Alzheimer's or heart-related conferences. Although a clear answer to determine this shortcoming may be difficult to make, what seems clear is that Alzheimer conferences are largely dominated by the pharmaceuticals that sponsor and fund these meetings and who are engaged in selling their drug products, in this case, anti-Abeta drug therapy. An example of this "arrangement" between pharmaceuticals and non-profit organizations is the Alzheimer's Association International Conference held in Washington, DC in 2015. This conference bills itself as "the largest forum for dementia research" and is publicly funded.

Looking at the program, over 80 % of the presentations and sessions dealt in one form or another with Abeta amyloid and issues related to this peptide. The major sponsors of the conference were the same pharmaceuticals, namely Eli Lilly and

Biogen, pandering to the attendees and to Wall Street investors whom they rely on to heavily bet on their their anti-Abeta drugs, which it should be said, have all failed despite elaborate and expensive clinical trials (see also Chap. 5). Ironically, there was no participant or attendee at this, the largest forum for dementia research, who challenged these sham proceedings. This could have been done by using reality as a template to describe that **no clinical progress, no help to patients with or without Alzheimer's, and no plan for the immediate prevention of this dementia have been realized or even attempted in decades due to the monomaniacal pursuit of a pie in the sky "pill" promoted by the pharmaceutical sponsors of the conference. This is a sad conclusion that seems to totally ignore the plight of those facing a disastrous disorder in favor of accommodating big pharma's out-of-control greed and indifference to human suffering.**

Excluding the Abeta myth, very few ideas and a chance to exchange or argue the merit of alternate research proposals were allowed by the Alzheimer's Association International Conference to filter into the think tank that forms the basis of a valid scientific forum. **Perhaps passionate love of the Abeta research is at fault. To quote H.L Mencken "The most costly of all follies is to believe passionately in the palpably not true."**

Young Alzheimer investigators and graduate students are the prime victims of this intellectual skewering since most are dependent on extracurricular funding to keep their jobs and their careers. The pernicious attitude of allowing one powerful paradigm, as the Abeta hypothesis has become, to dictate a common body of dubious assumptions, does no service to the scientific community. This disregard for scientific veracity is all the more disquieting when it is realized that the contradictions and inconsistencies exemplified by a string of uninterrupted clinical failures of the Abeta hypothesis cannot apparently be shaken from its foundation despite growing concern for its unassailable tenet [75–78].

Dr. Michael D'Andrea in his book **Bursting Neurons and Fading Memories** makes the point that considerable attention for the Abeta hypothesis has entirely depended on the removal of extracellular Abeta 42 plaques from the brain, but since this has not been achieved, the concept flounders because removal of those plaques by anti-Abeta therapy is not associated with reduced cognitive decline or neuronal loss [79].

Never stepping back from the possibility that the Abeta hypothesis appears faulty, the ardent supporters of this persuasion inexplicably conceive something to blame for the repeated clinical testing failures. The blame can fall on **not using the treatment early enough in the disease process, not using the proper drug dose**, or **not therapeutically targeting the most critical Abeta species** [80]. **In fact, almost any excuse will do as long as a new clinical trial can be spun and new stockholder shares can be sold**. Not once, have the chief subscribers of the Abeta hypothesis concluded from the negative clinical findings that the hypothesis may be on the wrong track and should be critically re-examined or abandoned altogether so that the field can move on to something more substantive. For academic investigators, such an admission would likely jeopardize their careers and reputations because if they come to the conclusion that Abeta is simply a pathological product that is deposited in Alzheimer brain tissue from other consequences and is not the cause of neuronal loss or cognitive decline, their funding could come to a dramatic end. It is a compelling conflict between professional survival and scientific integrity.

This state of affairs forces the conclusion that no useful progress in Alzheimer clinical research will be made as long as an unbending paradigm totally overshadows any competing ideas and possible remedies that could help patients dodge this dementia.

Thomas Kuhn in his brilliant book **Structure of Scientific Revolutions** [81] postulated that a bad paradigm will be replaced by a better paradigm only when enough significant anomalies have collected making the old paradigm obsolete and impractical. This outcome will earn the new paradigm credibility and an intellectual influence that may move science forward. Kuhn also notes that **when the faulty paradigm is replaced by a better one, the followers of the old paradigm will almost always clash intellectually with followers of the new paradigm until a much better paradigm replaces them both.**

One can see this evolutionary process of **paradigm shifts** in the short history of Alzheimer research. Initially, Alzheimer's disease was believed to occur as a dysfunction of the neurotransmitter acetylcholine which led to the characteristic neurodegeneration and cognitive decline. Scientists in the early 1980s noted that cholinergic neurons degenerated and died in the brain of Alzheimer patients and in those with mild cognitive impairment. They attributed this cholinergic abnormality to the initiation of cognitive decline observed in elderly people. The cholinergic hypothesis paradigm had an important influence on early Alzheimer drug development and led to a pharmaceutical bonanza as three cholinergic drugs (Razadyne, Exelon, and Aricept) were approved by the FDA for the treatment of Alzheimer's, creating a multi-billion-dollar market for big pharma. Some years later, it was discovered that assays on postmortem acetylcholine enzymes did not support the cholinergic hypothesis, a finding that led scientists to challenge not only the concept behind the paradigm but also the rationale for using cholinergic drugs to treat Alzheimer's, especially at the early stages of the disease.

These three cholinomimetic drugs are still prescribed for preclinical and clinical Alzheimer's, and many investigators still believe that they are neuroprotective. Their adverse side effects, notably nausea, vomiting, diarrhea, and bronchoconstriction, can, on a chronic basis, bring on other more lethal adverse events. **It is puzzling then, why they remain as prescriptive agents to treat mild and moderate Alzheimer's since evidence shows that these drugs have limited, if any, treatment effect** [82] **and may in fact accelerate Alzheimer's progression** [83]. This problem of Alzheimer patients paying for dubious medicines has made consumer protective groups question both the wisdom of the FDA and the way powerful drug lobbies influence Congressional legislators not to pass laws that protect the public and instead allow their "golden geese" to remain on the prescriptive market.

The fall of the cholinergic hypothesis left a gap in the paradigm thinking which was filled in 1992 by the Abeta hypothesis, discussed in Chap. 4. It might seem a coincidence that as the cholinergic and Abeta paradigms rose from clinical research studies, it was a pharmaceutical that stood to support either one and quickly elaborate therapy aimed at the presumed neuropathology. No banner, no cry of "hold your horses" before you begin aggressive therapy on patients is ever tooted in consideration of the scientific method and truth. This is rather disturbing because as scientists, we have a responsibility to pursue truth above all else and not put patients at risk of worsening their condition. That is how culture, technology, and civilized behavior are able to move forward. Investigators that are looking for the legitimacy of whether the Abeta paradigm is valid or not are not funded by pharmaceuticals and receive considerably less support by private, non-profit agencies dedicated to bettering the Alzheimer outlook since many of them exist only from pharmaceutical charity. Government funding of Alzheimer's research is even harder to obtain.

Another scientific discrepancy should be mentioned. Breast cancer in the USA affects about 23,200 people yearly, and the number of new cases has been steadily decline in the last 5 years due to improved therapy and medical prevention. Breast cancer research received $684 million in 2014 from the National Institutes of Health (NIH).

By contrast, Alzheimer's affects about 500,000 new cases every year and has an exponential growth of new cases expected to triple by 2050 (Fig. 9). Alzheimer's research received $562 million from NIH in 2014. This budgetary fiasco is not only incredibly stupid and shortsighted but is bound to cripple the budget allocated for Medicare and Medicaid within the next 35 years as the dementia population keeps growing unabated. Very little is done to convince the US Congress that less than $500 million for ALL Alzheimer-related research is spectacularly irresponsible and mindless.

The budget for Alzheimer's needs to be increased to at least $3 billion in fiscal year 2016 which is what HIV/AIDS receives from NIH despite being a highly preventable and treatable disease affecting one-tenth of the new Alzheimer cases diagnosed each year. This boost in funding for Alzheimer's research would better assist many young investigators starting their careers in neuroscience to opt for investigations that do not depend on cronyism and good old boys club among the NIH peer reviewers who determine the fate of each grant submission.

Alzheimer's disease is presently in a stranglehold of stagnant progress and offers little hope of helping patients avoid or stem the suffering from this dementia. The present focus of clinical research is to treat a nearly destroyed brain with a pill that will remove or block some bad proteins in the brain and keep the dementia patient in the grip of a zombie state, able to walk but not think. As tragic as this has become, few voices are raised to stop this mindless (although highly profitable) approach. **The abhorrent picture that emerges is that by treating a nearly destroyed brain, the patient will require more expensive pills, often for years before death mercifully ends the agony for both patient and caregiver**.

To shame the pharmaceutical top brass for playing this money game is pointless since they have shown no sense of shame or regret in the two decades that they have engaged in this particular practice. Instead, physicians and the public should denounce the practice and Alzheimer specialists recruited to carry out the clinical trials should refuse to do so on ethical principles and moral grounds. As long as the principal focus of clinical research is to find a pill to medicate people with irreversible and incurable dementia, no help to soften the impact of this disorder will occur.

In summary, good medical practice can prevent cognitive meltdown and dementia, but it requires a powerful interest and financial investment by the government and by more pragmatic non-profit agencies devoted to Alzheimer research and prevention, not those agencies dedicated to selling the fantasy of a "cure for Alzheimer's" (for details, see [84]).

The "cure Alzheimer's" fairy tale began in the 1990s when many non-profit agencies scrambled for donation money and by investigators who sought publicity from shoddy experiments. The "cure Alzheimer" delusion was regularly delivered to a misinformed public which eagerly wanted to see some hope for those struck by the disease. **This pandering for an Alzheimer cure continues to this day, aided largely by ignorant medical reporting in the lay press and by television media looking for breaking news stories.**

In my judgment, **we should no longer regard dementia as an inevitable disease of the aged**. This brief review indicates that cerebrovascular and cardiovascular disease can be prevented or treated using personalized interventions. **In** Chap. 18, **personalized interventions to prevent or delay severe cognitive impairment are discussed in detail.**

References

1. Safouris A, Psaltopoulou T, Sergentanis TN, Boutati E, Kapaki E, Tsivgoulis G. Vascular risk factors and Alzheimer's disease pathogenesis: are conventional pharmacological approaches protective for cognitive decline progression? CNS Neurol Disord Drug Targets. 2015;14 (2):257–69.
2. de la Torre JC. Hemodynamic consequences of deformed microvessels in the brain in Alzheimer's disease. Ann NY Acad Sci 1997;826:75–91
3. Lee SJ, Ritchie CS, Yaffe K, Stijacic Cenzer I, Barnes DE. A clinical index to predict progression from mild cognitive impairment to dementia due to Alzheimer's disease. PLoS ONE. 2014;9(12):e113535.
4. Thomas T, Miners S, Love S. Post-mortem assessment of **hypoperfusion** of cerebral cortex in **Alzheimer's disease** and vascular dementia. Brain. 2015;138(Pt 4):1059–69.
5. Xekardaki A, Rodriguez C, Montandon ML, Toma S, Tombeur E, Herrmann FR, Zekry D, Lovblad KO, Barkhof BF, Giannakopoulos P, Haller S. Arterial spin labeling may contribute to the prediction of cognitive deterioration in healthy elderly individuals. Radiology. 2014;2–14(274):490–99.
6. Alsop DC, Dai W, Grossman M, Detre JA. Arterial spin labeling blood flow MRI: its role in the early characterization of Alzheimer's disease. J Alzheimers Dis. 2010;20(3):871–80.
7. Qiu C. Preventing Alzheimer's disease by targeting vascular risk factors: hope and gap. J Alzheimers Dis. 2012;2012(32):721–31.
8. Bangen KJ, Nation DA, Delano-Wood L, Weissberger GH, Hansen LA, Galasko DR, Salmon DP, Bondi MW. Aggregate effects of vascular risk factors on cerebrovascular changes in autopsy-confirmed Alzheimer's disease. Alzheimers Dement. 2015;11(4):394–403.

9. Bangen KJ, Nation DA, Clark LR, Harmell AL, Wierenga CE, Dev SI, Delano-Wood L, Zlatar ZZ, Salmon DP, Liu TT, Bondi MW. Interactive effects of vascular risk burden and advanced age on cerebral blood flow. Front Aging Neurosci. 2014;6:159–60.
10. Clerici F, Caracciolo B, Cova I, Fusari Imperatori S, Maggiore L, Galimberti D, Scarpini E, Mariani C, Fratiglioni L. Does vascular burden contribute to the progression of mild cognitive impairment to dementia? Dement Geriatr Cogn Disord. 2012;34:235–43.
11. Mielke MM, Rosenberg PB, Tschanz J, Cook L, Corcoran C, Hayden KM, Norton M, Rabins PV, Green RC, Welsh-Bohmer KA, Breitner JC, Munger R, Lyketsos CG. Vascular factors predict rate of progression in Alzheimer disease. Neurology. 2007;69:1850–8.
12. Richard E, Moll van Charante EP, van Gool WA. Vascular risk factors as treatment target to prevent cognitive decline. J Alzheimers Dis. 2012;32:733–40.
13. Breteler MM. Vascular risk factors for Alzheimer's disease: an epidemiologic perspective. Neurobiol Aging. 2000;21(2):153–60.
14. de la Torre JC, Mussivand T. Can disturbed brain microcirculation cause Alzheimer' disease? Neurol Res. 1993;15:146–53.
15. de la Torre JC. Critically attained threshold of cerebral hypoperfusion: the CATCH hypothesis of Alzheimer's pathogenesis. Neurobiol Aging. 2000;21(2):331–42.
16. Chételat G, Desgranges B, de la Sayette V, Viader F, Eustache F, Baron JC. Mild cognitive impairment: can FDG-PET predict who is to rapidly convert to Alzheimer's disease? Neurology. 2003;60(8):1374–7.
17. Leary MC. Incidence of silent stroke in the US. stroke. Cerebrovasc Dis. 2003;16:280–5.
18. Omalu B. Chronic traumatic encephalopathy. Prog Neurol Surg. 2014;28:38–49.
19. Bouma GJ, Muizelaar JP. Cerebral blood flow in severe clinical head injury. New Horiz. 1995;3(3):384–94.
20. McKee AC, Cantu RC, Nowinski CJ, Hedley-Whyte ET, Gavett BE, Budson AE, Santini VE, Lee HS, Kubilus CA, Stern RA. Chronic traumatic encephalopathy in athletes: progressive tauopathy after repetitive head injury. J Neuropathol Exp Neurol. 2009;68(7):709–35.
21. Johnson NA, Jahng GH, Weiner MW, Miller BL, Chui HC, Jagust WJ, Gorno-Tempini ML, Schuff N. Pattern of cerebral hypoperfusion in Alzheimer disease and mild cognitive impairment measured with arterial spin-labeling MR imaging: initial experience. Radiology. 2005;234:851–9.
22. Xu W, Yu JT, Tan MS, Tan L. Cognitive reserve and Alzheimer's disease. Mol Neurobiol. 2015;51(1):187–208.
23. Stern Y. Cognitive reserve in ageing and Alzheimer's disease. Lancet Neurol. 2012;11 (11):1006–12.
24. Ott A, Breteler MM, van Harskamp F, Claus JJ, van der Cammen TJ, Grobbee DE, Hofman A. Prevalence of Alzheimer's disease and vascular dementia: association with education. The Rotterdam study. BMJ. 1995;310:970–3.
25. de la Torre JC. Cerebral perfusion enhancing interventions: a new strategy for the prevention of Alzheimer dementia. Brain Pathol, In press 2016.
26. Aaslid R. Visually evoked dynamic blood flow response of the human cerebral circulation. Stroke. 1987;18:771–5.
27. Roland PE, Friberg L. Localization of cortical areas activated by thinking. J Neurophysiol. 1985;53(5):1219–43.
28. Wilson RS, Segawa E, Boyle PA, Bennett DA. Influence of late-life cognitive activity on cognitive health. Neurology. 2012;78(15):1123–9.
29. Mazoyer B, Houdé O, Joliot M, Mellet E, Tzourio-Mazoyer N. Regional cerebral blood flow increases during wakeful rest following cognitive training. Brain Res Bull. 2009;80(3):133–8.
30. Colcombe S, Kramer AF. Fitness effects on the cognitive function of older adults: a meta-analytic study. Psychol Sci. 2003;14:125–30.
31. Kessler H, Taubner S, Buchheim A, Münte TF, Stasch M, Kächele H, Roth G, Heinecke A, Erhard P, Cierpka M, Wiswede D. Individualized and clinically derived stimuli activate limbic structures in depression: an fMRI study. PLoS ONE. 2011;6(1):e15712.

32. Bench CJ, Frackowiak RS, Dolan RJ. Changes in regional cerebral blood flow on recovery from depression. Psychol Med. 1995;25(2):247–61.
33. Cermakova P, Eriksdotter M, Lund LH, Winblad B, Religa P, Religa D. Heart failure and Alzheimer's disease. J Intern Med. 2015;277(4):406–25.
34. Roher AE, Tyas SL, Maarouf CL, Daugs ID, Kokjohn TA, Emmerling MR, Garami Z, Belohlavek M, Sabbagh MN, Sue LI, Beach TG. Intracranial atherosclerosis as a contributing factor to Alzheimer's disease dementia. Alzheimers Dement. 2011;7(4):436–44.
35. Rafnsson SB, Deary IJ, Fowkes FG. Peripheral arterial disease and cognitive function. Vasc Med. 2009;14(1):51–61.
36. Johnson W, Price JF, Rafnsson SB, Deary IJ, Fowkes FG. Ankle–brachial index predicts level of, but not change in, cognitive function: the Edinburgh Artery Study at the 15-year follow-up. Vasc Med. 2010;15(2):91–7.
37. Ito H, Kanno I, Fukuda H. Human cerebral circulation: positron emission tomography studies. Ann Nucl Med. 2005;19(2):65–74.
38. Rabkin SW. Arterial stiffness: detection and consequences in cognitive impairment and dementia of the elderly. J Alzheimers Dis. 2012;32(3):541–9.
39. LaRocca TJ, Hearon CM Jr, Henson GD, Seals DR. Mitochondrial quality control and age-associated arterial stiffening. Exp Gerontol. 2014;14(58C):78–82.
40. Anonymous. Cardiogenic dementia. Lancet;1977:309(8001):27–8.
41. de Toledo Ferraz Alves TC, Ferreira LK, Wajngarten M, Busatto GF. Cardiac disorders as risk factors for Alzheimer's disease. J Alzheimers Dis. 2010;20(3):749–63.
42. Zuccala G, Onder G, Marzetti E. Use of angiotensinconverting enzyme inhibitors and variations in cognitive performance among patients with heart failure. Eur Heart J. 2005;26:226–33.
43. Corder EH, Ervin JF, Lockhart E, Szymanski MH, Schmechel DE, Hulette CM. Cardiovascular damage in Alzheimer disease: autopsy findings from the Bryan ADRC. J Biomed Biotechnol. 2005;2005(2):189–97.
44. Boudoulas KD, Sparks EA, Rittgers SE, Wooley CF, Boudoulas H. Factors determining left atrial kinetic energy in patients with chronic mitral valve disease. Herz. 2003;28(5):437–44.
45. Thacker EL, McKnight B, Psaty BM, Longstreth WT Jr, Sitlani CM, Dublin S, Arnold AM, Fitzpatrick AL, Gottesman RF, Heckbert SR. Atrial fibrillation and cognitive decline: a longitudinal cohort study. Neurology. 2013;81(2):119–25.
46. Kähönen-Väre M, Brunni-Hakala S, Lindroos M, Pitkala K, Strandberg T, Tilvis R. Left ventricular hypertrophy and blood pressure as predictors of cognitive decline in old age. Aging Clin Exp Res. 2004;16(2):147–52.
47. Clarke RJ. Homocysteine-lowering trials for prevention of cardiovascular events: a review of the design and power of the large randomized trials. Am Heart J. 2006;151:282–7.
48. Harris E, Macpherson H, Pipingas A. Improved blood biomarkers but no cognitive effects from 16 weeks of multivitamin supplementation in healthy older adults. Nutrients. 2015;7 (5):3796–812.
49. McCully KS. Vascular pathology of homocysteinemia: implications for the pathogenesis of arteriosclerosis. Am J Pathol. 1969;56:111–28.
50. Hofman A, Ott A, Breteler MM, Bots ML, Slooter AJ, van Harskamp F, van Duijn CN, Van Broeckhoven C, Grobbee DE. Atherosclerosis, apolipoprotein E, and prevalence of dementia and Alzheimer's disease in the Rotterdam Study. Lancet. 1997;349(9046):151–4.
51. Lai WK, Kan MY. Homocysteine-induced endothelial dysfunction. Ann Nutr Metab. 2015;67 (1):1–12.
52. Hoth KF, Poppas A, Moser DJ, Paul RH, Cohen RA. Cardiac dysfunction and cognition in older adults with heart failure. Cogn Behav Neurol. 2008;21(2):65–72.
53. Waldstein SR, Giggey PP, Thayer JF, Zonderman AB. Nonlinear relations of blood pressure to cognitive function: the Baltimore longitudinal study of aging. Hypertension. 2005;45:374–9.
54. Forette F, Seux ML, Staessen JA, Thijs L, Babarskiene MR, Babeanu S, Bossini A, Fagard R, Gil-Extremera B, Laks T, Kobalava Z, Sarti C, Tuomilehto J, Vanhanen H, Webster J, Yodfat Y, Birkenhäger WH. Systolic hypertension in Europe investigators. The prevention of

dementia with antihypertensive treatment: new evidence from the systolic hypertension in Europe (Syst-Eur) study. Arch Intern Med. 2002;162:2046–52.
55. Bosch J, Yusuf S, Pogue J. HOPE investigators. Heart outcomes prevention evaluation. Use of ramipril in preventing stroke: double blind randomised trial. BMJ. 2002;324:699–702.
56. Tzourio C, Anderson C, Chapman N. PROGRESS collaborative group. Effects of blood pressure lowering with perindopril and indapamide therapy on dementia and cognitive decline in patients with cerebrovascular disease. Arch Intern Med. 2003;163:1069–75.
57. Skoog I, Lithell H, Hansson L. SCOPE study group. Effect of baseline cognitive function and antihypertensive treatment on cognitive and cardiovascular outcomes: study on cognition and prognosis in the elderly (SCOPE). Am J Hypertens. 2005;18:1052–9.
58. Pak T, Cadet P, Mantione KJ, Stefano GB. Morphine via nitric oxide modulates beta-amyloid metabolism: a novel protective mechanism for Alzheimer's disease. Med Sci Monit. 2005;11 (10):BR357–66.
59. Channon KM, Qian H, George SE. Nitric oxide synthase in atherosclerosis and vascular injury: insights from experimental gene therapy. Arterioscler Thromb Vasc Biol. 2000;20 (8):1873–81.
60. Pearson TA, Mensah GA, Alexander RW, Anderson JL, Cannon RO 3rd, Criqui M, Fadl YY, Fortmann SP, Hong Y, Myers GL, Rifai N, Smith SC Jr, Taubert K, Tracy RP, Vinicor F. Centers for disease control and prevention; American Heart Association. Markers of inflammation and cardiovascular disease: application to clinical and public health practice: a statement for healthcare professionals from the Centers for Disease Control and Prevention and the American Heart Association. Circulation. 2003;107(3):499–511.
61. Galleano M, Pechanova O, Fraga CG. Hypertension, nitric oxide, oxidants, and dietary plant polyphenols. Curr Pharm Biotechnol. 2010;11(8):837–48.
62. Breteler MMB, Claus JJ, Grobbee DE, Hofman A. Cardiovascular disease and distribution of cognitive function in elderly people: the Rotterdam Study. BMJ. 1994;308:1604–8.
63. Cuadrado-Tejedor M, Hervias I, Ricobaraza A, Puerta E, Pérez-Roldán JM, García-Barroso C, Franco R, Aguirre N, García-Osta A. Sildenafil restores cognitive function without affecting β-amyloid burden in a mouse model of Alzheimer's disease. Br J Pharmacol. 2011;164 (8):2029–41.
64. Cukierman T, Gerstein HC, Williamson JD. Cognitive decline and dementia in diabetes–systematic overview of prospective observational studies. Diabetologia. 2005;48(12):2460–9.
65. Davies RS, Vohra RK, Bradbury AW, Adam DJ. The impact of hormone replacement therapy on the pathophysiology of peripheral arterial disease. Eur J Vasc Endovasc Surg. 2007;34 (5):569–75.
66. Hagger-Johnson G, Sabia S, Brunner EJ, Shipley M, Bobak M, Marmot M, Kivimaki M, Singh-Manoux A. Combined impact of smoking and heavy alcohol use on cognitive decline in early old age: Whitehall II prospective cohort study. Br J Psychiatry. 2013;203(2):120–5.
67. Ford AB, Mefrouche Z, Friedland RP, Debanne SM. Smoking and cognitive impairment: a population-based study. J Am Geriatr Soc. 1996;44(8):905–9.
68. Sabia S, Elbaz A, Dugravot A, Head J, Shipley M, Hagger-Johnson G, Kivimaki M, Singh-Manoux A. Impact of smoking on cognitive decline in early old age: the Whitehall II cohort study. Arch Gen Psychiatry. 2012;69(6):627–35.
69. Polidori MC, Schulz RJ. Nutritional contributions to dementia prevention: main issues on antioxidant micronutrients. Genes Nutr. 2014;9(2):382.
70. Kesse-Guyot E, Andreeva VA, Jeandel C, Ferry M, Hercberg S, Galan P. A healthy dietary pattern at midlife is associated with subsequent cognitive performance. J Nutr. 2012;142 (5):909–15.
71. Ngandu T, Lehtisalo J, Levälahti E, Laatikainen T, Lindström J, Peltonen M, Solomon A, Ahtiluoto S, Antikainen R, Hänninen T, Jula A, Mangialasche F, Paajanen T, Pajala S, Rauramaa R, Strandberg T, Tuomilehto J, Soininen H, Kivipelto M. Recruitment and baseline characteristics of participants in the finnish geriatric intervention study to prevent cognitive impairment and disability (FINGER)-a randomized controlled lifestyle trial. Int J Environ Res Public Health. 2014;11(9):9345–60.

72. Poggesi A, Salvadori E, Valenti R, Nannucci S, Ciolli L, Pescini F, Pasi M, Fierini F, Donnini I, Marini S, Chiti G, Rinnoci V, Inzitari D, Pantoni L. The florence VAS-COG clinic: a model for the care of patients with cognitive and behavioral disturbances consequent to cerebrovascular diseases. J Alzheimers Dis. 2014;42(Suppl 4):S453–61.
73. van Buchem MA, Biessels GJ, Brunner la Rocca HP, de Craen TJM, van der Flier WM, Ikram MA, Kappelle J, Koudstaal PJ, Mooijaart SP, Niessen W, van Oostenbrugge R, de Roos A, van Rossum B, Daemen MJAP. The heart-brain connection: a multidisciplinary approach targeting a missing link in the pathophysiology of vascular cognitive impairment. J Alzheimers Dis 2014;42(Suppl 4):S443-51.
74. Zea-Sevilla MA, Fernández-Blázquez MA, Calero M, Bermejo-Velasco P, Rábano A. Combined Alzheimer's disease and cerebrovascular staging explains advanced dementia cognition. Alzheimers Dement. 2015 (in press).
75. Morris GP, Clark IA, Vissel B. Inconsistencies and controversies surrounding the amyloid hypothesis of Alzheimer's disease. Acta Neuropathol Commun. 2014;18(2):135.
76. Robakis NK. Are **Abeta** and its derivatives causative agents or innocent bystanders in AD? Neurodegener Dis. 2010;7(1–3):32–7.
77. Giannakopoulos P, Herrmann FR, Bussiére T, Bouras C, Kövari E, Perl DP, Morrison JH, Gold G, Hof PR. Tangle and neuron numbers, but not amyloid load, predict cognitive status in Alzheimer's disease. Neurology. 2003;60:1495–500.
78. Doody RS, Farlow M, Aisen PS; Alzheimer's disease cooperative study data analysis and Publication Committee. Phase 3 trials of solanezumab and bapineuzumab for Alzheimer's disease N Engl J Med. 2014 10;370(15):1460.
79. D'Andrea MR. Bursting neurons and fading memories. Academic Press; 2015, pp. 27–32.
80. Liu E, Schmidt ME, Margolin R, Sperling R, Koeppe R, Mason NS, Klunk WE, Mathis CA, Salloway S, Fox NC, Hill DL, Les AS, Collins P, Gregg KM, Di J, Lu Y, Tudor IC, Wyman BT, Booth K, Broome S, Yuen E, Grundman M, Brashear HR. Amyloid-β 11C-PiB-PET imaging results from 2 randomized bapineuzumab phase 3 AD trials. Bapineuzumab 301 and 302 clinical trial investigators. Neurology. 2015;85(8):692–700.
81. Kuhn TS. The structure of scientific revolutions. 3rd ed. Chicago, Illinois: The University of Chicago Press; 1996.
82. Birks JS, Grimley Evans J. Rivastigmine for Alzheimer's disease. Cochrane Database Syst Rev. 2015;4:CD00119.
83. Parnetti L, Chiasserini D, Andreasson U, et al. Changes in CSF acetyl- and butyrylcholinesterase activity after long-term treatment with AChE inhibitors in Alzheimer's disease. Acta Neurol Scand. 2011;124:122–9.
84. de la Torre JC. Alzheimer's disease is incurable but preventable. J Alzheimers Dis. 2010;20 (3):861–70.

Chapter 13
The Good, the Bad, and the Ugly of Advanced Aging

Good Aging

A recent news item buried on page 12 of the Daily Mail in London tells of a 75-year-old British nurse who travelled to Basel, Switzerland, seeking assisted suicide which is legal in that country. The news story would not be news if the explanation for this lady's choice of ending her life had not been so bizarre. Although this nurse was in relatively good health for her age, it turns out that her former occupation was in geriatric care and she believed she was better off dead than eventually becoming one of her frail elderly patients. This story is an extreme view of how many people view their so-called golden years, a stage of life loosely connected with retirement from active work. Very little can be drawn from an event where a healthy, elderly individual wishes to die, especially in this case since the nurse in question probably had access to scheduled drugs that could have easily ended her life without having to travel to Switzerland. But therein may lie the clue.

While it is the dream of many people to welcome retirement age as a form of mental and physical freedom not felt since childhood, the short-lived reward of doing what you want and not be bound by work-related responsibilities may not help keep you alive and healthy. Many studies have shown that elderly people in retirement who engage in sedentary life, who do not exercise mind or body and who overeat, are many times at risk of developing cognitive decline and subsequent progression to Alzheimer's. Recommendations from **human prevention trials point to good nutrition, physical exercise such as aerobic and muscle strength training, cycling, walking for an hour every day, mental activities that require attentive thinking, social activities, and intensive monitoring by a qualified physician of potential or actual vascular risk factors as important modifiers not only for future mental deterioration but also for a healthier mind and body reaching the sixth to the tenth decade of life** [1].

Readers of these recommendations might well ask: "who's got the time?" The simple answer is that the time taken to stay healthy is an investment in staying alive

J.C. de la Torre, *Alzheimer's Turning Point*,
DOI 10.1007/978-3-319-34057-9_13

longer and enjoying the rewards of that investment. The rewards of good health during aging cannot be overemphasized since they allow a good quality of life during a period of mental and physical adjustment.

The human accomplishments made during advanced aging are legendary. **Isaac Newton** worked productively in highly important mathematical concepts until age 85; **Winston Churchill** completed a six-volume account of World War II at age 79; **Goethe**, Germany's greatest writer, produced his masterpiece *Faust*, at age 82; **Harriet Tubman** freed more than 300 slaves through the Underground Railroad and founded a house for the poor in her 80s; **Miguel de Cervantes**, Spain's greatest writer completed Don Quijote in his 1970s; **Michaelangelo** finished the Sistine Chapel in the Vatican between the ages of 72–89; **Victor Hugo** wrote some of his greatest novels after age 75; **Verdi** composed the opera Falstaff at 80 and sacred musical works at age 85; **Benjamin Franklin** represented the USA in France brilliantly from age 76 to 79; and **Titian** painted great masterworks after age 90 and died at 99 while working on the *Pieta*. These examples are but a fraction of the achievements made by people after age 60 and constitute an auspicious reminder that life is not over… until it is.

The scientific evidence concerning the effect of exercise on strengthening muscles and keeping the brain functional as long as possible, particularly in the elderly, is convincingly vast.

As we age, muscles tend to become weaker and may undergo withering from lack of use. Muscle weakness may occur as we age possibly because of lower blood flow or some unknown physiochemical abnormality. Added to this, older people are more inactive and sedentary and many suffer from vitamin D deficiency which can lead to muscle atrophy, particularly in the lower limbs. Weakness of lower limbs can lead to difficulty rising from a chair or climbing stairs. But, most serious of all, are falls that cause bone fractures (see next section, **Falls**).

Vitamin D deficiency is common in the elderly due to less sunlight exposure, reduced dietary intake, and reduced intestinal absorption for some vitamins and nutrients. When tests show low blood levels of vitamin D, supplements of this vitamin can improve muscle strength in many patients.

Age-related loss of skeletal muscle is called sarcopenia. Lean muscle mass loss in sarcopenia is estimated at 3–8 % in individuals over age 60 and 50 % loss after age 80. Although research continues to find the mechanisms involved in this muscle mass loss, no cause has yet been found. What is known is that sarcopenia is the cause of marked disability in the aging person.

Dr. Charles Ambrose, a professor of immunology at the University of Kentucky, has published an interesting explanation. Dr. Ambrose presents evidence that a reduction of capillaries in the elderly muscle tissue contributes to muscle weakness as a result of diminished angiogenesis [2]. Angiogenesis is a physiological process involving the formation of new blood vessels from preexisting vessels. It is a vital process for the growth and development of an organism and is activated during wound repair and from muscle contractions. If Dr. Ambrose is correct in his proposal about a deficit in angiogenesis causing sarcopenia, it may well explain the reduced muscle contractions experienced during advanced aging which are believed to occur from faulty neurotransmission. Hence, Dr. Ambrose proposes that angiogenic factors should be considered as therapeutic targets to stimulate angiogenesis much like thyroxin or testosterone are given to patients with these hormonal deficiencies [2]. If successful, such therapy would revolutionize the management of age-related muscle weakness and frailty. Regular exercise is also able to release angiogenic factors activated by muscular movements and could explain the salutary effects of this activity in lowering the risk of Alzheimer's during aging [2].

Hypertension is another major challenge during advance aging. About two-thirds of US adults over age 60 have hypertension, usually isolated systolic hypertension and a normal diastolic pressure. Blood pressure is given in two numbers, systolic pressure which is the top number and diastolic pressure. Systolic pressure refers to the blood being pumped out for each heartbeat as it moves through the arteries and the diastolic pressure is the time the heart relaxes and fills with blood.

Optimal Blood Pressure

There has been an ongoing debate among the experts as to **what constitutes an optimal blood pressure for people over the age of 60**. Years ago, the rule of thumb formula 100-plus-age and subtracting 10 for women was considered a normal blood pressure which did not require treatment. This meant that a 70-year-old with a blood pressure of 170 was thought to have a satisfactory blood pressure. However, this concept was proven false and replaced with newer guidelines that recommended blood pressure of 120 over 80 mm Hg **regardless of age**.

The thinking became that a normal pressure for a 20-year-old, that is, 120 over 80, should be the target for the elderly as well. That ideal pressure did not take into consideration that during advanced aging, arterial stiffening and increased vascular resistance to the heart's pumping force requires a higher pressure to adequately supply blood to the brain and other organs.

Until now, older people were told to strive for blood pressures below 140/90, with many taking multiple drugs to achieve that goal. More recently, a panel of experts representing **The Joint National Committee on Blood Pressure Management for Adults eased previous guidelines for people over age 60 without diabetes or chronic kidney disease and recommended that the goal for the systolic pressure could be below 150 mm Hg rather than 140 and the diastolic below 90 before starting treatment** [3]. This expert panel concluded that there was no compelling evidence that previous guidelines for a target blood pressure lower than 140 systolic served any useful or medical purpose and that, in addition, drugs used to achieve target blood pressures below 140 systolic had associated risks for adverse events in the elderly person. These new guidelines have not been totally accepted by many physicians who argue that more lenient blood pressure targets for the elderly might lead to dangerous higher blood pressure being ignored by primary care physicians when seeing patients. The debate about when to start antihypertensive treatment has not addressed what optimal blood pressure level should be the target for people already receiving antihypertensives. A 6-year study of over 300 people age 75–101 reported that **a higher incidence of dementia was seen in people whose systolic blood pressure was over 180 compared to those with systolic blood pressures 140 and below** [4]. The same study showed evidence that the risk of dementia also increased when diastolic pressure fell below 65.

The report by **The Joint National Committee on Blood Pressure Management for Adults** recommending that antihypertensive therapy did not need to be started in people aged 60 and over with systolic pressures under 150 was recently challenged by the **Sprint Report**, a National Institutes of Health study on 9300 men and women over age 50. Sprint advocates that people at high risk of heart disease or kidney disease should aim to lower their systolic blood pressure to less than 120. This systolic pressure target is lower than any other previous guideline has ever suggested. Proof of concept for the low systolic pressure was given by observing a reduced risk of heart attack and stroke by nearly 33 % and death by nearly 25 % in individuals whose systolic pressures were kept below 120. However, the report does not mention how many elderly individuals acquired cognitive impairment and dementia while keeping their systolic pressures below 120. The medical paper containing these findings is scheduled to be published in 2016, and until the report is peer-reviewed and minutely examined, these findings should be regarded with caution.

One quick test that can detect whether vascular reactivity is healthily working in the elderly individual is to check whether during sleep, the systolic blood pressure dips below 10 % from daytime blood pressure. If it does not, vascular reactivity may be impaired. This simple test can be done with an unobtrusive blood pressure monitor that an awake person can record. Vascular reactivity is described as the ability of smooth muscle cells to either contract or relax following an environmental vasoconstricting or vasorelaxant stimulus. The inability to respond appropriately to such stimuli can lead to harmful blood pressure changes, which among other things can affect the cerebral vasculature and cognitive function.

Apart from treating hypertension, a good way to age well is to combine physical with mental activities. This active life can begin at any age but is acutely more essential after retirement from work. The sedentary lifestyle that can be blamed for increasing the chance of early cognitive difficulties may be compensated when mental activities are engaged such as reading, playing word games, writing, attending lectures, and seeking laughter in movies, books, magazines, and countless other pastimes that force one to use memory and reasoning. However, if not combined with physical activities, the protective edge afforded by intellectual past-times against cognitive dysfunction is less likely to be as effective.

Higher educational and professional attainment has been widely reported to lower the risk of cognitive decline and of Alzheimer's when compared with people with limited education. This social disparity has been attributed to what several investigators call better "**cognitive reserve**" by the group with higher education. Thus, according to this notion, a person finishing college or better is less likely to develop cognitive decline in later life than an individual with less education and occupational attainment. Epidemiological studies have confirmed this social disparity when two groups with different educational backgrounds are compared, but these studies have not determined what lies behind this incongruity [5, 6].

Cognitive Reserve

Some investigators have suggested that "cognitive reserve" in the elderly offers some degree of protection from age-related neuropathology. Simply stated, cognitive reserve has been postulated as a brain with a higher threshold for damage, particularly when that damage targets regions of the brain concerned with cognitive function. People who age with a high cognitive reserve, according to this idea, will show more resilience to neuronal damage during aging. Higher education and professional attainment are thought to be independent contributors to cognitive reserve.

Despite a surging interest in attempting to explain why some people appear more immune to dementia symptoms or even have extensive plaque deposits in their brains but show no cognitive disability, the concept of cognitive reserve has gained a loyal following by some investigators due to suggestive findings that allow this conclusion to be considered. However, there is no solid evidence that cognitive reserve exists under its present definition and other contrary findings have questioned this idea [7, 8].

We have argued that what seems to be important in the resilience of the brain to mildly protect against age-related wear and tear damage is not a high IQ or professional attainment but rather the ability to engage in frequent mental activities that can induce a boost in cerebral blood flow [9]. **These mental activities are bound to be used by people with higher education who are more prone to engage in intellectual pursuits and by high achievers. This partly explains the differences seen in cognitive testing for cognitive resilience; however, in my judgment, this difference is not exclusive to this socially privileged group**.

Excluding the $APOE_4$ genotype from the two populations who differ in the level of education [10], a low-educated or unskilled worker with intellectual curiosity can engage in critical thinking and achieve the same degree of protection from age-related cognitive decline. For example, Wilson and colleagues [11] evaluated over 1000 cognitively normal elderly individuals with a mean age of 80 years and **diverse levels of education and professional attainments**. After a 5-year follow-up, it was found that those individuals who engaged in sustained, mentally stimulating activities such as playing chess or word puzzles, reading newspapers or books, and visiting a library, regardless of educational attainment or professional backgrounds, predicted better cognitive function than those who did not engage in such mental activities. This study received support from another study that showed lower educated people who engaged in frequent mentally stimulating activities could achieve the same level of protection against memory loss as better educated subjects [12].

This disparity between higher and lower education and relative risk to Alzheimer's has nevertheless baffled scientists who have offered various theories, including cognitive reserve, to explain the phenomenon. However, one clue that may help explain the high-low education–Alzheimer's risk ambiguity is the generally accepted correlation between neurovascular coupling and cognitive function. Consequently, an increase in metabolic demand, such as from critical thinking, demands an increase in brain blood flow [13]. Following this reasoning, studies have shown that mental activity requiring certain types of thinking can increase oxidative metabolism and regional cerebral perfusion in multiple cortical and subcortical fields including the hippocampus, one of the initial sites of Alzheimer pathology [14].

The resultant average increase in cerebral blood flow following mental stimulating activities is about 10 % from baseline [11], and appears to be temporary. However, if performed on a routine basis, it may explain what is believed to be cognitive reserve in some individuals (regardless of their educational attainment) who engage in such mental activities. For these reasons, *cognitive preserve* rather than "cognitive reserve" may be a more accurate description for people who engage in stimulating mental activities that appear to protect the brain against cognitive decline. This cognitive protection seems to depend more on what a person of advanced age does than what that person did in the distant past. Cognitive protection solely due to professional attainment, IQ or higher education may be a spurious inference based on a positive correlation that ignores other variables.

Evidence has been shown in Chap. 1 (see mixed dementia) how cerebral blood flow naturally diminishes as we age as a result of acquiring risk factors that generally appear during advanced aging. People with intellectual curiosity and the desire to pursue mentally stimulating activities may offset to a degree the natural fall in cerebral blood flow which is estimated to drop 1.0 % every two years or about 15–20 % from age 20 to 60. This age-related reduction in cerebral blood flow which can result in cognitive dysfunction during advanced aging and its apparent interventional delay by mental–physical activation is important to recognize because it may provide a form of cognitive reserve which appear more of a "cognitive advantage" than a "reserve." Thus, the ability to dynamize cerebral perfusion during advanced aging by engaging in mental and physical activities in addition to being treated when Alzheimer's risk factors are present is a more cogent explanation why such people have a lower risk of developing Alzheimer's. The lifestyle favoring mental–physical activity to delay dementia onset has been reported in numerous clinical studies. For these reasons, "cognitive preserve" rather than "cognitive reserve" may be a more accurate description for people who engage in stimulating mental activities that appear to protect the brain against cognitive decline. This cognitive protection seems to depend more on what a person of advanced age does than what that person did in the distant past. Cognitive protection solely due to professional attainment, IQ or higher education may be a spurious inference based on a positive correlation that ignores other variables. This contentious issue is further discussed in Chap. 16.

From the evidence presented, it is reasonable to speculate that all things being equal, an uneducated but mentally active individual has the same opportunity of lessening the risk of Alzheimer's as an Oxford don.

Stress in Life

Just as muscle workouts tend to strengthen muscles in the body, **managing and actively preventing injurious stress** will help the brain cope with environmental aggravations laced with life-threatening consequences.

What is stress? The word stress is now part of the common language and may have different definitions and meanings for different people. A more physiological definition points to stress as a complex physiological reaction to stressors.

The renowned endocrinologist **Dr. Hans Selye** coined the term *stress* in 1936 in his landmark book *The Stress of Life, From Dream to Discovery* [15]. Selye's definition of stress was brief and called it "the non-specific response of the body to any demand for change." Selye revised this definition later in his life so that the public could associate body change to any demand as "the rate of wear and tear on the body."

These definitions are rarely used today even in conversations among scientists and instead, stress has become a buzzword for anything bad that happens to the body where no easy explanation can be found. Selye, however, spent much of his lifetime studying stress mechanisms and concluded that, without stress, a person would be a vegetable or dead [15].

Thus, the thought of a "good" and "bad" stress was born from Selye's work on the role of hormones, particularly corticosteroids and their effect on the brain. Harmful stress was shown by Selye to be involved in many disease states including high blood pressure, stroke, coronary thrombosis, and arterial stiffening. Since these conditions are presently consistent with vascular risk factors to Alzheimer's, it is no wonder that "bad" stress has also been linked to neurodegeneration and dementia. The "good" part of stress (also called eustress) is that it can motivate people to accomplish more and be more productive in their life. **The roller coaster analogy has been used to explain "good' and "bad" stress**. Some people who board the roller coaster for the first time sit in the back of the coaster, and with each steep plunge, they shut their eyes, clench their teeth, and hold white-knuckled to the retaining bar hoping the stressful ride will soon be over; others sit in the front of the coaster, yell gleefully, relish each steep dive, and cannot wait to repeat the experience.

Harmful stress is usually continuous but can also be acute. Either type of stress tends to take a toll on concentration, memory, and decision-making activities that are under the control of the prefrontal cortex. The prefrontal cortex makes extensive connection with other cortical and subcortical brain regions that regulate thought, action, and emotion. This brain region is also responsible for detecting incorrect errors and providing the insight to correct them using new strategies, a task called self-control errors.

The prefrontal cortex is also the brain area that is most sensitive to the damage that acute episodes or repeated stress exposure can cause. When the prefrontal cortex is attacked by stressors, self-control errors that are mediated by prefrontal cortex can be suppressed leading to maladaptive behaviors involving overeating, drug addiction, alcoholism, and depression. Personal vulnerability to the consequences of advanced aging can often be diminished by engaging in activities that keep the brain and body functional.

Many self-control errors, maladaptive behaviors, adverse life events, and humdrum of life in adults can be managed preferably by removing the stressor whenever possible. This may be possible with the counseling help of a medical expert, for example, psychopharmacologists, behavioral neuroscientists, social and personality psychologists, among others. Stressors can also be managed or removed by the person undergoing the stress, but this requires recognition of the stressors and formulating a strategy to control them. **Listening to soft music or pleasant nature sounds, socializing with friends, practicing yoga and meditation, acquiring a fun hobby, avoiding lack of sleep, and exercising daily are only a few of the activities older people have successfully tried to relieve stress. The point here is to keep busy as you age with activities that will not overload the brain and the body but will keep them flexible and engaged**.

Unfortunately, there are stressors that cannot be easily diminished or removed, and these will be discussed below in the section titled "The bad" of advanced aging.

An excellent reflection of good aging is reaching age 100 or beyond with a good quality of life. These centenarians now number about 80,000 in the US, and their numbers are growing.

An ongoing study of centenarians by The New England Centenarian Study at Boston University School of Medicine is funded by the National Institutes of Health. The study was begun in 1996 with the purpose of discovering what factors lead to healthy longevity. Below is a summary of their findings.

- Few centenarians are obese. In the case of men, they are nearly always lean.
- Substantial smoking history is rare.
- A preliminary study suggests that centenarians are better able to handle stress than the majority of people.
- Centenarians ($\sim$15 %) had no significant changes in their thinking abilities. This finding disproves the expectation that all centenarians would be demented. It was also discovered that Alzheimer's disease was not inevitable. Some centenarians had very healthy appearing brains following neuropathological examination after death (we call these gold standards of disease-free aging).
- Many centenarian women have a history of bearing children after the age 35 and even 40 years. From our studies, a woman who naturally has a child after the age of 40 has a 4 times greater chance of living to 100 compared to women who do not. It is probably not the act of bearing a child in one's forties that promotes long life, but rather doing so may be an indicator that the woman's reproductive

system is aging slowly and that the rest of her body is as well. Such slow aging and the avoidance or delay of diseases that adversely impact reproduction would bode well for the woman's subsequent ability to achieve very old age.
- At least 50 % of centenarians have first-degree relatives and/or grandparents who also achieve very old age, and many have exceptionally old siblings. Male siblings of centenarians have an 17 times greater chance than other men born around the same time of reaching age 100, and female siblings have an 8½ greater chance than other females also born around the same time of achieving age 100.
- Many of the children of centenarians (age range of 65–82 years) appear to be following in their parents' footsteps with marked delays in cardiovascular disease, diabetes, and overall mortality.
- Some families demonstrate incredible clustering for exceptional longevity that cannot be due to chance and must be due to familial factors that members of these families have in common.
- Based upon standardized personality testing, the offspring of centenarians, compared to population norms, scores low in neuroticism and high in extroversion.
- Genetic variation plays a very strong role in exceptional longevity.
- People who reach 100 and beyond can be classified into 3 groups, survivors, delayers, and escapers. However, we found that many of our centenarian subjects had age-related diseases even before the age of 80 (about 43 %, and whom we called "survivors"), after the age of 80 (about 42 % and whom we called "delayers"), and lastly, those who had no mortality-associated diseases at age 100 (about 15 % and whom we called escapers).
- Eighty-five percent of centenarians are women and 15 % are men. Among supercentenarians, the female prevalence may increase to about 90 %. Though women by far win the longevity marathon, paradoxically, fewer men are generally functionally better off and healthier. This may be because women handle age-related diseases better (how they do this is not clear), whereas at these ages, the men more readily die from them. Thus, the men who survive have to be relatively healthy and functionally fit.

The New England Centenarian Study at Boston University School of Medicine continues its research into how "good aging" may be achieved, a topic that is intimately relevant to avoiding dementing disorders, notably Alzheimer's disease.

Bad Aging

Alzheimer's Has no Cure

The most terrifying aspect of being accurately diagnosed with Alzheimer's dementia is that it is not only a progressive, unrelenting, devastating disorder

to mind and body, but most significantly, there is no cure for it. Although Alzheimer's incurability is a harsh and sad conclusion, there are inescapable biological facts in support of it.

One of the noblest aspects of practicing medicine is that professionals can provide hope to many patients whose illnesses were difficult or impossible to treat 50 years ago. But, there are limits to what medicine can do for some diseases now or in the future. Alzheimer's disease is a perfect example of this limitation.

When cognitive functions such as memory, reasoning, language and planning appear to be lost during the evolution of Alzheimer's, it is because the neuronal networks and their connections (which prior to dementia controlled the lost cognitive abilities) are likely dead or fatally damaged and **dead brain cells cannot be brought back to life**. Replacement of degenerated brain cells in **Alzheimer's, assuming it were technically feasible, is unlikely to be achieved without creating a new personal identity in the host, in other words, a new human being**.

To better understand what the problem of a cure for Alzheimer's is all about, the definition of what is a "cure" is necessary. An "incurable disease" has been variously defined as a malady whose cure is impossible or incapable of being corrected, for reasons that define the malady.

This definition appears both fatalistic and historically inaccurate since a less rigid definition should leave open the possibility that many "incurable diseases" are simply those in which a cure has not yet been found. For example, a century ago, cancers such as acute lymphocytic leukemia in children, Burkitt's lymphoma, Wilm's tumor, Hodgkin's disease, rhabdomyosarcoma, testicular cancer, osteogenic sarcoma, were regarded as "incurable." Nonetheless, advances in selective surgery, radiation therapy, and chemotherapy or a combination of these procedures have made it possible to reasonably "cure" or put into remission these cancers in many people if early intervention is applied.

Other "incurable" and deadly diseases now considered potentially curable when intervention is applied at an early pathological stage are bubonic plague, aneurysm of the aorta, malaria, syphilis, tuberculosis, bacterial pneumonia, peptic ulcer, toxic shock syndrome, subdural hematoma, vital organ transplants for dysfunctional heart, liver, kidney, pancreas, lung or bone marrow, and many others. A century ago, many of these disorders were considered "incurable."

It should be noted that at the present time, most disorders affecting the nervous system are considered "incurable." **I maintain that the term "incurable" has been misapplied to many neurological disorders where we simply have not found a way to successfully reverse the symptoms and outlook**. Typical examples of neural diseases deemed incurable are amyotrophic lateral sclerosis, Alpers disease, Brown-Séquard syndrome, Charcot–Marie–Tooth disease, and Huntington's chorea, to name only a few. As mentioned above, historical proof has repeatedly discredited the strict definition that an "incurable" morbid process cannot be cured at its onset, even when neural damage has occurred. Does that infer that most neurological diseases are potentially curable? The answer is a qualified yes, especially when the culprit agent responsible for the disease has been identified. Are there some neurological diseases that are firmly and irrevocably incurable? The answer is yes again, as discussed below. Brain cell degeneration and neuronal death are the underlying features of most neurological diseases. Until now, embryonic nigral cell implants, which have shown clinical benefit in some patients for the treatment of Parkinson's disease, have been the only attempt at replacing degenerated neurons.

In my judgment, it is a matter of intuitive logic supported by an empirical confidence in scientific innovation to predict that practically all sensory–motor neuron diseases now considered incurable will someday be manageable or even curable.

Conceivably, fatally damaged motor, sensory, and brainstem neurons and their connections which mostly control body movement, sensation, or vital life functions should, theoretically at least, be amenable to replacement with viable working neurons, neurogenesis, or stem cell technology so that a lifeless body function can be restored. Such replacement of fatally damaged or dead motor, sensory, and brainstem neurons can be conceivably transplanted to successfully rewire lost somatic function. However, that type of sensory–motor neuron replacement technology has not yet been achieved and may never be. **The glowing exception to curing most neurological disorders that spare mental competence are diseases such as dementia that destroy the brain cells that govern the intellectual qualities which make us who we are** [16].

Leaving aside Aristotelian sophistry concerning brain–mind rhetoric, it is not the same thing to replace neurons whose main function is to deliver signals to the body to move or feel as it is to replace neurons that regulate cognitive function.

Dementias, such as Alzheimer's, Niemann–Pick's disease, vascular dementia, and Korsakoff's syndrome, among others, fall into this category and can be described as **firmly incurable disorders once cognitive-related brain cells have perished. These dementias are not to be confused with temporary "dementing" conditions (not true dementias) that are often treatable when caused by depression, drugs, alcohol, or metabolic imbalance since these can generally be reversed if the provoking factor is identified and corrected.**

To clearly understand what is required to replace dead or dying neurons that govern cognitive abilities, a brief review of how Alzheimer's begins and ends may be useful (see also Chap. 1).

The chronology of Alzheimer's development follows three main stages of increasing severity according to the Clinical Dementia Rating (CDR). These stages are described as mild, moderate, and severe. Mild- or early-stage Alzheimer's is characterized by recent memory loss, personality changes, poor judgment, language-related problems, depression, and apathy.

Neuropathologically, the mildest form of this stage (CDR = 0.5) already shows a 60 % loss of neurons in the entorhinal cortex, as compared to cognitively normal controls. The entorhinal cortex is a critical region that forms an integral component of the medial temporal lobe memory system and is commonly damaged at the early stages of Alzheimer's [17].

In the middle-stage of Alzheimer's dementia (CDR = 2), daily assistance in order to cope with the deterioration of most cognitive functions is required.

Persistent memory loss and behavioral problems worsen considerably over a span of 8–10 years at this stage. Progressive neuronal loss in the entorhinal cortex, hippocampus, and other territories in the limbic system and cerebral cortex is the characteristic of this stage.

The final and severe stage of AD (CDR = 3) is marked by total loss of recognition of people and places, general incapacitation, and inability to care for oneself, requiring full-time 24-7 daily care. Death may follow after several years from a number of causes due to a severely weakened and vulnerable physical state. Pathologically, the number of neurons in layers in the entorhinal cortex further decrease by about 70–90 % together with marked shrinkage of the brain [17] (Fig. 1.4).

It is hard to envision how any improvement could be secured in the face of progressive cognitive deterioration and brain degeneration at any of the three Alzheimer's stages. Such a feat would need to involve replacing the dead or degenerating brain cells responsible for the acquired cognitive deficits, a task that, even if it were technically possible, would involve the creation of a new human being.

This is the main reason for classifying neurodegenerative dementias that lay waste to the mind as firmly incurable. **Consequently, replacing neuronal networks and their connections which prior to their death was charged with regulating emotions, memory, abstraction, reasoning, language, and other cognitive functions that define the individual cannot be done without creating a new personal identity**. Such a process cannot be considered a "cure" to a person so afflicted despite its paradoxical outcome.

Falls in the Elderly

On average, 33 % of elderly people experience at least one fall per year [18]. The number of elderly patients visiting the emergency department with ground-level falls is rising due to the increasing age in the general population. It is estimated by the Centers for Disease Control and Prevention that in one year, 2.5 million nonfatal falls among older adults are treated in emergency departments with and more than 700,000 of these patients being hospitalized. Many falls in the elderly involve traumatic brain injury which is a common cause of death in the older person. Rates of fall-related bone fractures among older women are twice that of men.

The best predictor of falling during advanced aging is a previous fall. There are many reasons why elderly people fall. For example, medication that dulls the senses or judgment, frailty, bad eyesight, environmental factors such as a slippery surface, unfamiliarity with a new terrain, impaired balance, muscle weakness (sarcopenia), and sudden dizziness on rising from a chair is only a few of the many reasons given for falling. Falls in the elderly differ somewhat from younger people who tend to fall from someone bumping into them, doing athletic activities, or tripping with environmental obstacles. Many elderly adults are frail, their bones are more brittle, and they have preexisting medical conditions that can be more difficult to manage even after a low-level fall.

The gravity of older people falling is the consequences of the fall. Aside from the physical damage resulting from a fall in an elderly individual, about half of these elders are unable to get up without help. Lying on the ground for hours can bring about dehydration, hypothermia, or pneumonia. Ambulation may deteriorate from fracturing a hip or brittle leg bone. Loss of mobility may seriously affect quality of life, and fear of falling again may add to a greater need for assisted living and less personal independence.

Surviving older living is a major challenge. Aside from dementia, an older person is more likely to face certain disorders such as hypertension, cardiac arrhythmias, type 2 diabetes, visual disturbances such as macular degeneration and glaucoma, pain from osteoarthritis or joint disease, many different cancers, psychiatric conditions, and specific attack on every organ of the body.

It is indeed a minor triumph to run through this unforgiving medical gauntlet and not be seriously or fatally affected by them until the eighth or ninth decade of life. Modern medicine has helped extend quality of living by providing interventions and care not available 50 years ago. The elderly individual can also add to the help that medicine has provided by using common sense and by self-education on how to best survive old age. The "bad" part of aging can be considerably softened by following the general recommendations provided in the "good" part of aging. This requires that a shift in thinking be made from middle age to advancing age viewing one's youth as a vanished epoch that will never return. The trauma, infections, stress, and other insults to mind and body flung at us at youth and most often tolerated by a more resilient mind and body are to be avoided by overriding common sense as old age approaches; this mind-set needs to focus on prevention of harm rather than the *laissez faire* attitude of youth.

So far as this section on aging is concerned, we have only briefly explored how the passage of time affects the brain and ways that may protect our intellect and emotions. From the neck down, the readers are on their own.

The Ugly

Contemplating Suicide

The subject of suicide is never pleasant to write about. Religious, moral, ethical, and medical issues complicate any discussion about suicide. With respect to Alzheimer's in the USA, it is a reality that needs to be discussed due to the booming increase in the elderly population and the dramatic explosion of new Alzheimer's cases which are estimated to skyrocket from a present 5 million to 14 million by 2050. Alzheimer's is presently ranked as the sixth leading cause of death in the USA with about 700,000 people dying with this disorder every year.

The projected increase of Alzheimer's dementia by 2050 is appalling. Aside from the misery shared by victims of this dementia and their caretakers, Alzheimer's is expected to consume more that 30 % of the Medicare budget at an annual cost of over 1 trillion dollars annually.

An understated and publicly restrained aspect of acquiring Alzheimer's is the patient's consideration of suicide or physician assisted suicide as a way to avoid the relentless and destructive power of dementia that dehumanizes mind and body. With respect to the exponential growth of new Alzheimer's cases, suicide is a reality that needs to be publicly discussed.

A blogger named **David Hilfiker** wrote this:

> Many people with Alzheimer's lose their *ability* to (consider) suicide before they're ready to go through with it. The novel *Still Alice* depicts an intellectual college professor with early-onset Alzheimer's who decides she'll suicide when it gets 'too bad'. She places into the medicine chest a bottle of pills strong enough to kill her and writes herself a note with exact instructions on when and how to use the pills. As she gradually declines, she gradually loses the capacity to remember or figure out what she meant. Ultimately, she wonders who this person is who is trying to kill her by writing this note to her.

The case of a 65-year-old geriatric nurse, who travelled to Switzerland where assisted suicide is lawful, is discussed in the section above on "good" aging. This nurse did not appear to suffer from any life-threatening disease or debilitating pain, and she merely wanted to avoid becoming as frail and confused as some of her patients, so she ended her life. This unconventional case of a healthy person seeking death on the premise that further aging will turn ugly defies the imagination, but conceivably, the decision may have been nurtured by an underlying depression or some other change in nurse's mental state. However, bizarre as this case has become, it is motivationally far removed from the extreme situation facing Alice in the novel *Still Alice* who knows she has entered **the realm of "cognitive effacement"** where awareness of self and the external world slowly vanishes forever.

A real case of Alzheimer-related suicide surfaced in the news in 2014, about Sandra Bem who five years before had been diagnosed with dementia at the age of 65. Here is an excerpt from that story written by Alix Spiegel for Northwest Public Radio (NPR):

> Sandra Bem wasn't a stranger to suicide. She and her husband Daryl Bem, were both psychologists, professors at Cornell University in Ithaca, N.Y. Both had volunteered at a suicide hot line and so had an intimate appreciation of just how destructive the act of suicide could be. Typically, in the wake of suicide, family member are devastated. Rates of depression in those left behind are much higher than when a loved one dies a natural death.

The story of Sandra Bem ends 5 years later at the age of 70 when declining cognitive function is markedly impaired but not to the point where she does not know what to do. She asks and is given the drugs she has selected over the years to overdose on. Her family members grieve, but they are spared seeing the mental indignities and body breakdown that Alzheimer's will keep on heaping on Sandra Bem for many years. Like others before her, Sandra Bem could have opted for physician assisted suicide, but she apparently preferred preparing her family for this decision several years before, and in the end, she appeared stoically prepared as she was surrounded by them on the day of departure.

The subject of suicide is rarely talked about in the news media or in public forums even though it has been debated since ancient times. In the *Republic*, and in *Laws*, **Plato** argued that treatment to prolong the life of severely ill patients should be withheld because it would create a burden to themselves and others. Plato did not regard suicide as morally wrong in some cases, for example, when it was compelled by extreme unavoidable misfortune. His pupil Aristotle did not agree with his mentor on this issue.

We know what ancient Greeks thought about suicide when they forced Socrates to drink a fluid-containing hemlock for the alleged crime of corrupting the youth and ironically, for failing to acknowledge the teachings of the gods.

Platonic thinking was echoed by the Roman Stoic philosopher **Seneca**, who wrote in 4 AD that an individual should have broad discretion to end his or her own life. The sixteenth-century philosopher, **Michel de Montaigne**, proposed that suicide was not a question of Christian belief but a matter of personal choice. In an essay, Montaigne argued both sides of the suicide issue and concluded that suicide was an acceptable moral choice depending on the circumstances, adding that "pain and the fear of a worse death seem to me the most excusable incitements."

St. Thomas Aquinas in his Summa Theologica argued that suicide was unlawful and a sin against God because life is God's gift to man. God alone, according to Aquinas, decides when life shall begin and end. However, the Bible does not mention suicide directly as a sin, but it does state in Luke 12:10 that God will forgive all sins except blasphemy against the Holy Spirit.

A counter argument to Aquinas thinking is that, if there is a God, no one knows what "He" wants. People interpret what God wants through religious writings and from the ecclesiastic or religious representatives of their particular faith or from personal belief independent of religion. Alzheimer's dementia was not known when these sacred writings were prepared so it must be assumed that God's wishes, as such, are interpretable.

Throughout the ages, many have opposed the idea of suicide or assisted suicide because it constitutes killing. This view has a long history within the context of diverse religious, philosophical, and personal beliefs.

The arguments against suicide generally revolve around religious teaching with the central theme being that only God should choose whether we live or die. Christians believe that by suffering they get closer to God and share the suffering of Christ. Many conservative Protestants argue that to commit suicide is self-murder that is similar to murdering someone else.

Jews believe that suicide or assisted suicide is a serious sin and thus is forbidden by Jewish law. However, there is nothing in the Talmud that prohibits suicide. Historically, Jews have committed mass suicide to avoid slavery as in Masada, where in 73 CE, 960 rebels and their families chose to die by self-immolation as the Roman troops stormed the fortified garrison. However, in the post-Talmudic tractate *Semachot,* which deals with people on their deathbeds, burials, and mourning, it is stated that those who commit suicide will not receive burial rites.

Islam also views suicide as a grave sin as instructed in the Quran, and most Muslim scholars extend this dictum to Hamas and jihad worshippers who are motivated to become suicide bombers. The paradox within Islam is that terrorist suicide fighters must not kill noncombatants such as children, women, and the old, but martyrdom is guaranteed for those who die fighting non-Muslims. This rule is ignored by Muslim terrorists who strap bombs to their chest and kill other Muslims and anyone else who stands in their way, whether they are noncombatants, women or children.

Buddhism unlike other major religions does not condemn suicide but concludes that this act is bound to have negative consequences on the path to enlightening and upon rebirth. Despite the Buddhist lack of rejection for suicide, it in no way advocates any destruction of life.

In the USA, only 5 states as of 2015 allow assisted suicide, Washington, Montana, Oregon, Vermont, and New Mexico. In Washington and Oregon, the laws allow terminally ill adults to acquire lethal doses of medicines from their physicians, while the other 3 states have the right to seek assisted suicide or lethal medication as long as a provided protocol is followed. Many other states are putting pressure on their legislators to approve death-with-dignity laws, but it is unclear whether Alzheimer's disease is considered a terminally ill disease before severe cognitive decline sets in. Other countries besides Switzerland now approve assisted suicide and include Germany, the Netherlands, Belgium, Luxembourg, Albania Japan, Colombia, and the Canadian province of Quebec.

In the USA, about 40,000 suicide deaths are recorded every year, and the rates are 4 times higher for men than for women. Since there is still much social stigma to suicide, it is hard to estimate how many people with Alzheimer's choose this option either from assisted suicide (which most states do not allow to be recorded in the death certificate by a physician) or by self-deliverance.

Derek Humphry in his book *Final Exit* [19] discusses the option of suicide or assisted suicide written for people who face or may face a fatal illness. In his book, Humphry advises ingesting a lethal quantity of secobarbital sodium capsules (Seconal Sodium) by pouring the contents of the capsules into a half glass of water and rapidly drinking the dissolved barbituric salts. Humphry writes about the Hollywood myth that has created an almost idyllic form of death and dying. He writes:

> We have become so brain-washed by the fast, usually bloodless, and always painless deaths shown continually by the movie and television production industry that our collective perceptions of the act of death are sanitized. Whether by gunshot or through illness, the actor just rolls over and that's the end. We want so much to believe that this is true that we don't question it.

The tormented path from Alzheimer's to death has to be one of the most dreaded of all diseases known to mankind. There is a consensus among experts in the field that the use of feeding tubes, cardiac pulmonary resuscitation (CPR), and heroic measures to keep an advanced Alzheimer's patient alive is of little benefit and likely adds to the suffering of the patient. This ugly conversion generally begins as a person slowly transitions from a relatively healthy and mentally sound state to an insidious and progressive loss of memory, intellect, self-esteem, and independence.

Soon, emotional turmoil in the form of irritability, anger, depression, and rage become the daily pattern, compounded by a lack of motor skills such as talking, walking, or swallowing, eventually resulting in the inability to control bladder and bowel movements causing feces to leak from the rectum, being incapable or unaware how to provide self-care, lacking communication skills, needing help with daily activities much like a recently born infant but, unlike an infant, unable to recognize and interact with close family members or the environment. Death from infection, pneumonia, severe dehydration, or heart attack may finally come 8–12 years; many will be bedridden and tube-fed for much of that time. At the final stage of Alzheimer's, it is very difficult to predict when the patient will die as it would be for a patient with cancer. Medical technology can keep a patient alive in advanced Alzheimer's for a long time, in some ways prolonging death rather than quality of life.

Alzheimer's is terribly democratic when it comes to radiating the distress and misery that it causes not only to the patient but to the caretakers, family and friends. The option of whether to choose an early death after serious Alzheimer's symptoms begin to appear or to ride the course of the disease to its bitter end is a choice to make.

The will to live is a powerful force. Witness the millions of Nazi captives held in concentration camps during WW II who were starved, tortured, and abused on a daily basis and who had very little hope of surviving their ordeal. Few prisoners attempted to escape because they knew it meant a certain death when caught, and fewer still flung themselves on the electrified wire fences that surrounded the camps. Many of the prisoner deaths in those camps were written up as "suicides" or "attempted escape" by the SS Death's Battalion guards who "justified" their random killing of prisoners for any of a number of infractions. Despite their woes, hope to stay alive as long as possible was in the minds of most prisoners even when death was their constant companion.

It is understandable for a patient with mild Alzheimer's to want to keep on living. But what happens when reason and planning skills have disappeared as it will happen in Alzheimer's and the choice of dying has been erased from the mind? This is a matter to be discussed with close family members before dementia symptoms make this task impossible. It should be recalled that **no one has ever recovered from Alzheimer's or any other dementia and death without such recovery is a medical certainty**. Ugly as this is, we can ponder about what we wish to do if and when Alzheimer's makes its grim appearance. Whatever the choice, suicide is a topic that needs to be urgently discussed as a public health problem with respect to Alzheimer's and other dementias, and it benefits no one to ignore or sweep the problem under the rug.

Consider the following theoretical scenario. A magic fairy appears before an Alzheimer's patient who shows no recognition of family, environment, or self-awareness for months. The fairy magically taps the patient's forehead allowing 30 min of full cognition and a brief video to show the patient's behavior in the last two weeks. The fairy then asks "do you wish to continue as you are or end life right now?" The answer will be a key to how life and death are individually perceived while we remain cognizant and are able to make rational decisions but that power is lost as the mind slips away. Since there is no magic fairy to help make that decision, the informed reader may want to enlist family or close friends to discuss the issue and reveal his or her wishes in the event Alzheimer's strikes. Alzheimer's should not be regarded as other terminal diseases such as cancer or AIDS where rational decisions can be made about continuing life to the very end or terminate the suffering by suicide. When everything that constitutes life such as cognition and awareness has vanished, there is no life, only a shell of what used to be**. Nathanial Hawthorne's** meditative wish does not always materialize, "**Death should take me**," he said, "**while I am in the mood**."

References

1. Andrieu S, Coley N, Lovestone S, Aisen PS, Vellas B. Prevention of sporadic Alzheimer's disease: lessons learned from clinical trials and future directions. Lancet Neurol. 2015. pii: S1474-4422(15)00153-2.
2. Ambrose C. Muscle weakness during aging: a deficiency state involving declining angiogenesis. Ageing Res Rev. 2015. pii: S1568-1637(15)00033-1.
3. James PA, Oparil S, Carter BL, Cushman WC, Dennison-Himmelfarb C, Handler J, Lackland DT, LeFevre ML, MacKenzie TD, Ogedegbe O, Smith SC Jr, Svetkey LP, Taler SJ, Townsend RR, Wright JT Jr, Narva AS, Ortiz E. 2014 evidence-based guideline for the management of high blood pressure in adults: report from the panel members appointed to the eighth joint national committee (JNC 8). JAMA. 2014;311(5):507–20.
4. Qiu C, von Strauss E, Fastbom J, Winblad B, Fratiglioni L. Low blood pressure and risk of dementia in the Kungsholmen project: a 6-year follow-up study. Arch Neurol. 2003;60(2):223–8.
5. Sando SB, Melquist S, Cannon A, Hutton M, Sletvold O, Saltvedt I, White LR, Lydersen S, Aasly J. Risk-reducing effect of education in Alzheimer's disease. Int J of Geriatr Psychiatry. 2008;23:1156–62.
6. Kliegel M, Zimprich D, Rott C. Life-long intellectual activities mediate the predictive effect of early education on cognitive impairment in centenarians: a retrospective study. Aging Ment Health. 2004;8:430–7.
7. Christensen H, Hofer SM, Mackinnon AJ, Korten AE, Jorm AF, Henderson AS. Age is no kinder to the better educated: absence of an association investigated using latent growth techniques in a community sample. Psychol Med. 2001;31(1):15–28.
8. Mackinnon A, Christensen H, Hofer SM, Korten A, Jorm AF. Use it and still lose it? The association between activity and cognitive performance established using latent growth techniques in a community sample. Aging Neuropsychol Cognition. 2003;10:215–29.
9. de la Torre JC. Is Alzheimer's disease a neurodegenerative or a vascular disorder? Data, dogma, and dialectics. Lancet Neurol. 2004;3(3):184–90.
10. Vemuri P, Lesnick TG, Przybelski SA, Machulda M, Knopman DS, Mielke MM, Roberts RO, Geda YE, Rocca WA, Petersen RC, Jack CR Jr. Association of lifetime intellectual enrichment with cognitive decline in the older population. JAMA Neurol. 2014;71(8):1017–24.
11. Wilson RS, Segawa E, Boyle PA, Bennett DA. Influence of late-life cognitive activity on cognitive health. Neurology. 2012;78(15):1123–9.
12. Lachman ME, Agrigoroaei S, Murphy C, Tun PA. Frequent cognitive activity compensates for education differences in episodic memory. Am J Geriatr Psychiatry. 2010;18(1):4–10.
13. Aaslid R. Visually evoked dynamic blood flow response of the human cerebral circulation. Stroke. 1987;18:771–5.
14. Roland PE, Eriksson L, Stone-Elander S, Widen L. Does mental activity change the oxidative metabolism of the brain? J Neurosci. 1987;7(8):2373–89.
15. Selye H. The Stress of Life. New York: McGraw-Hill; 1956.
16. de la Torre JC. Alzheimer's disease is incurable but preventable. J Alzheimers Dis. 2010;20:861–70.
17. Gomez-Isla T, Price JL, McKeel DW Jr, Morris JC, Growdon JH, Hyman BT. Profound loss of layer II entorhinal cortex neurons occurs in very mild Alzheimer's disease. J Neurosci. 1996;16:4491–500.
18. Grafman J, Salazar AM. The ebb and flow of traumatic brain injury research. Handb Clin Neurol. 2015;128:795–802.
19. Humphry D. FinalExit. 3rd ed. New York: Dell Publishing; 2002.

Chapter 14
A Personal Account of How a Scientific Hypothesis Blooms into a Life of Its Own

The Barf Bag Theory

It was Monday and I was headed for the airport. That weekend had been spent sweating out the data from six previous experiments where **long-term brain hypoperfusion** had been induced in **young and aged rats**. My own brain needed a rest and I was looking forward to sleeping for a few hours on the flight from Montreal to Paris where I was scheduled to give a talk the next day.

The previous year had been quite busy in my Neurosurgery Laboratory at the University of Ottawa. I had just settled in for the 7-h flight and was gathering my thoughts over the talk I was going to give to a group at the Cerebrovascular Research Unit of the Centre Nationale de la Recherche Scientifique (CNRS), the largest government research organization in France and in Europe.

The talk in Paris was about how we had developed a rat model of persistent and progressive memory loss by permanently tying-off two or three of the four blood vessels that supply blood flow to the brain. We had seen how permanently tying-off these blood vessels in rats lowered their brain blood flow by about 40 %. Carotid artery ligation is done in humans but only one vessel is involved for a brief duration. The reason for this procedure in people is to surgically remove plaque material from inside the carotid artery that is blocking vital blood flow to the brain. It is also performed in people who are diagnosed with an internal carotid artery aneurysm in order to resect the portion of the artery that is involved in the aneurysm. Cognitive testing can be performed in these patients before and after carotid artery ligation to make sure the procedure has not permanently impaired memory or cognitive function [1].

In one of our studies, we tested memory recall and learning ability after carotid artery ligation on two groups of rats, young and old. Ligated rats were observed and tested for nine weeks. Following surgical recovery, ligated old rats appeared physically normal, but after a few weeks, they lost their exploratory behavior or curiosity when placed in an open field environment. This is an unusual behavior

J.C. de la Torre, *Alzheimer's Turning Point*,
DOI 10.1007/978-3-319-34057-9_14

because rats are very inquisitive of new environments and will generally sniff novel objects in their search for food. When placed in a large pool of cloudy water called the Morris Water Maze, rats are trained, prior to surgery, to find a slightly submerged platform by using visual cues placed around the pool to help them navigate toward the platform and escape the water, which they hate (Fig. 14.1). The Morris Water Maze is a widely use technique in behavioral neuroscience that tests learning and memory. It is a very useful marker of hippocampal damage (where short-term memory is received) and hippocampal function. In our tests, old rats, whose common carotid arteries had been ligated, could remember how to find the submerged platform and escape by using the visual cues placed around the pool (Fig. 14.1). These rats were removed from the water pool, dried, and returned to their cages if they failed to find the submerged platform within 120 s.

Old rats without carotid artery ligation can find the platform in about 10–15 s after training, while those with carotid artery ligation take 90 s or more to find the platform despite the same training. The water tank test can be performed before and after surgical ligation of the carotid arteries. We were surprised to see that after 9 weeks, recovery of lost memory was regained spontaneously without

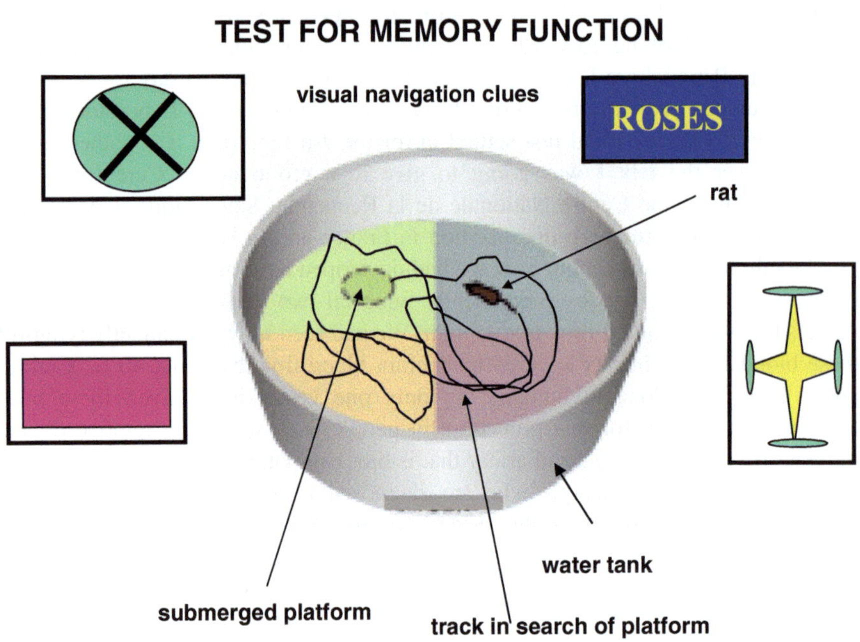

Fig. 14.1 Water tank used to test rat's ability to remember a previous trained task involving finding a slightly submerged escape platform placed in one of the four quadrants of the tank. Aged control rats are able to find the platform by using visual cues placed around the tank within 15 s. After carotid artery ligation reducing cerebral blood flow by 34 %, rats require 60–120 s to find the platform and their performance deteriorates progressively with time. Rats are removed from the water tank if they can not find the platform within 120 s. This test is repeated every 3–4 weeks until the end of the study

any treatment in the young but not the old rats. In fact, the old rats' memory and recall worsened with time, and after 6 months, atrophic changes of the brain began to emerge [2, 3] (Fig. 14.1). It was clear that lowering blood flow to the brain in old rats was inducing irreparable cognitive, neurochemical, and structural damage in the brain. **This finding had never been reported before because prior to us, no one, to my knowledge, had ever created a rat model of chronic brain hypoperfusion in any animal, particularly to study behavior** [4].

We carried out a series of experiments which confirmed and extended these findings [2–4]. This aging rat model of carotid artery ligation that we developed proved to be of fundamental value for research on cognition and aging because no drugs, brain lesions, or genetic manipulation were used to elicit memory loss and learning disability. Instead, damage to cognitive-related brain regions was achieved outside the brain by physiologically dwindling cerebral blood flow delivery to brain cells over short and long periods of time. This approach mimicked a number of chronic cardiovascular diseases and peripheral artery disorders common in aging patients who experience gradual and progressive memory loss.

Our experiments corroborated our suspicion concerning the relationship between chronic brain hypoperfusion during aging and progressive cognitive decline. Cognitive decline in these rats was associated with progressive neurodegeneration, neurochemical and molecular changes in the brain tissue, but curiously, these changes in the old rats occurred in the absence of Abeta-containing plaques. **These findings implied that Abeta was not necessary to create progressive neurodegeneration or cognitive impairment despite the severe nature of the brain damage created by unrelenting brain hypoperfusion**.

Several major findings from these experiments strongly suggested that aging rats subjected to lowered brain blood flow (but not age-matched controls or young rats) appeared to develop an Alzheimer-like syndrome. We saw no evidence that brain hypoperfusion caused strokes or ischemic lesions in the white matter which would have indicated that vascular dementia had occurred. Unlike vascular dementia which usually arises suddenly from blocked brain vessels, the cognitive symptoms of chronic brain hypoperfusion are manifested over weeks or months and the brain damage over 6 months (Fig. 4.1).

Since we were the first to introduce carotid artery ligation in rats to examine the role of chronic brain hypoperfusion on memory function in 1992 [2–4], we had a considerable jumpstart from the rest of the field to examine a number of variables relevant to how Alzheimer's may develop.

We found for example:

1. When old rats were compared to young rats receiving carotid artery ligation, memory loss continued and worsened only in the old rats; young rats spontaneously recovered their memory after 3 weeks even when their carotid arteries remained closed. This finding suggested that **advanced aging has a greater vulnerability to reduced cerebral blood flow** which is reflected by impaired cognition much like we see in Alzheimer's patients.

2. Old rats with carotid artery ligation showed not only worsening of their memory ability with time but also developed **severe brain atrophy of the cortex and hippocampus after 6 months** (Figure 4.1). This outcome simulated the neurodegenerative and atrophic process that characterizes Alzheimer's structural pathology (Fig. 1.4).
3. Old rats showed a 34 % reduction in cerebral blood flow shortly after carotid artery ligation much like the drop in brain blood flow seen in patients with mild cognitive impairment prior to converting to Alzheimer's dementia.
4. Old rats with carotid artery ligation showed elevation of a marker (called G protein) which is also a marker of early Alzheimer's.
5. Other pathological brain changes in old rats subjected to carotid artery ligation included oxidative stress, metabolic slow down, reduced acetylcholine levels, overgrowth of astrocytes, capillary distortions, lowered protein synthesis, and death of many hippocampal and cortical neurons. These changes reflected the pathological changes seen in humans prior to or after the onset of Alzheimer's disease [2–4].
6. Old rats whose memory was severely reduced after 2 weeks of carotid artery ligation regained their memory following the restoration of cerebral blood flow by microsurgical anastomoses of the occluded carotids [5].

These findings from the rat experiments suggested we were on the brink of understanding the true trigger of how Alzheimer's develops from brain hypoperfusion, but critical parts of the puzzle would remain unresolved until 1997.

The Vascular Hypothesis of Alzheimer's Disease Is Born

When these results were gathered and closely examined, **we could tentatively conclude that reducing cerebral blood flow in old rats but not young rats led to a cognitive and pathological state resembling Alzheimer's disease**. This rat-to-human correlation had significant implications. Although the rat evidence was suggestive, going from a rodent brain to a human brain can be a leap of faith that possibly ends in scientific blunder and embarrassment. Sitting in that flight to Paris in January, 1992, a piece of the Alzheimer's puzzle began to unfold before my eyes and I realized that the rat studies had provided intriguing evidence that was difficult to dismiss. **It was what is sometimes laughingly called a eureka moment.** The evidence from dozens of rat experiments seemed to be screaming at me. I looked into my briefcase with excitement to find paper to write on but somehow I had missed packing my yellow pad that I always used on trips to write down ideas and routine stuff I needed to do at a later time. I searched the seat pocket for anything that looked like paper and only found a "barf bag," euphemistically called airsickness bag.

This will do, I thought. I wrote: "After a series of experiments using aged rats subjected to various time periods of cerebral hypoperfusion, there appears a possibility that Alzheimer's disease is a vascular disorder with neurodegenerative consequences, not the other way around." A cascade of thoughts and ideas soon filled both sides of the barf bag and I had to call the flight attendant to ask for two more bags which she gingerly handed me, all the while looking at me suspiciously to make sure she could jump out of the way in case the lunch they had served would suddenly decorate her perfectly neat, blue uniform.

On my return from Paris, I could hardly wait to get back to the laboratory to put all the stats, ideas, and findings together. After an exhaustive search of the scientific literature, it appeared that several clinical researchers had lightly touched on the subject of cerebral blood flow and psychiatric conditions including dementia [6, 7], **but to my amazement, no one had apparently made the obvious connection that low brain blood flow during aging was the initiator of cognitive decline and the development of Alzheimer's dementia**.

An article written in 1976, by **Drs. Bo Hagberg** and **David Ingvar**, pioneers in blood flow studies of the human brain, reported reduced cerebral blood flow in specific regions of patients with *presenile dementia*, a condition now called familial Alzheimer's disease whose cause is due to a genetic mutation and whose incidence is much less common than sporadic Alzheimer's [8]. **These two investigators attributed the reduced cerebral blood flow they had observed in their demented patients to a loss of neurons and degenerative brain damage**. This explanation has survived to the present time and has been adopted by the authoritative **Diagnostic and Statistical Manual of Mental Disorders IV** (DSM-IV), a primary guide used by clinicians to diagnose mental conditions, including Alzheimer's disease. The DSM-IV considers that Alzheimer's can only be diagnosed after death when brain examination reveals the presence of Abeta-containing plaques and neurofibrillary tangles. The problem with this definition is that many cognitively normal-aged individuals show abundant Abeta plaques and neurofibrillary tangles whether living or dead [9–11], and conversely many patients diagnosed with Alzheimer's show few or no Abeta plaques [12].

Such findings imply that Abeta deposits in the brain and neurofibrillary tangles are either preclinical markers of impending Alzheimer's or random pathologic debris secondary to neurodegenerative disease. Since many brains of patients who die with heavy Abeta deposition but no indication of cognitive difficulties [9] as shown by prior psychometric testing, the second possibility seems more compelling.

With respect to Hagberg and Ingvar's [8] observation that reduced cerebral blood flow in familial Alzheimer's patients was due to degenerative changes, it was recently shown using state-of-the-art positron emission tomography (PET) and arterial spin labeling MRI that a parallel association between brain hypoperfusion and hypometabolism exists in brain regions where familial Alzheimer's damage is greatest [13]. But since no PET or MRI brain scans were available in 1976, it seemed logical but incorrect to conclude that the marked neurodegeneration observed histologically in dementia brains was responsible for the structural abnormalities noted in the microvasculature.

This conclusion was dumfounding to me, but then again, these researchers did not have our rat data of induced chronic brain hypoperfusion which might have provided the opposite conclusion. Nevertheless, the thought that a number of brilliant minds could not make the vascular–neurodegeneration connection, while we could, was exhilarating. The task of convincing our peers of this concept now lay ahead. To do this, I decided in 1992 to collate the basic laboratory rat findings together with how I pictured the clinical and the hemodynamic pathology would clash to create Alzheimer's disease. This dementia was, I firmly believed, primarily a vascular disorder with neurodegenerative consequences. This thinking was heresy and intolerably ignorant to the leading Alzheimer experts of the day. I also chose to report these findings in November, 1992, at the largest neuroscience meeting in the world, the Society for Neuroscience [14]. My presentation at the Neuroscience meeting was, as I anticipated, virtually ignored by the attendees. One attendee who seemed somewhat interested in my presentation was **Dr. Raj Kalaria**, an Alzheimer's researcher from Case Western Reserve University who remains intrigued with the vascular concept to this day.

Now I thought it was time to publish the basic and clinical evidence by submitting a very detailed paper with my colleague **Dr. Tofy Mussivand**, a cardiovascular physiologist who helped me understand the intricacies of cardiac hemodynamics. An important part of my education into the complex field of cerebral hemodynamics was to read again and again and again, the man many consider the father of biomechanics, **Professor Yuang-Cheng Fung's** book on biomechanics [15]. This text is an excellent treatise explaining basic and practical rules of how blood or Newtonian fluids, such as water, behave in extensible blood vessels or rigid tubes.

In my paper with Tofy, the experimental and clinical evidence linking Alzheimer's to poor blood flow to the brain was put together, and the article was submitted to two top-ranked journals that publish a lot of papers on Alzheimer's disease. Both journals rejected our paper, one reviewer calling it "contrary to accepted knowledge." I was glad not to have mentioned in the rejected article that the whole idea of calling Alzheimer's a vascular disorder originated in a barf bag. Of course, our technicians and close colleagues thought a good name was appropriate for this vascular proposal and jokingly called it the "barf bag theory."

Another neuroscience journal finally accepted our paper [16] in 1993, and the vascular hypothesis of Alzheimer's disease made its debut. Our paper had taken 6 months to collate the data and 3 months to write the results but the irony is that it was roundly ignored after its publication. Ignoring a scientific idea that may change how a disease is treated or managed is quite common in medicine and in science. A case in point is the 1677 discovery by a Dutch fabric merchant named Anton van Leewenhoek who built the first powerful microscope and reported "animalcules" (later renamed microbes) in a paper published by the science journal Philosophical Transactions. These animalcules came from Leewenhoek's own mouth and from polluted fluids and implied that microscopic "bugs" were in our bodies, a fact which was ignored for nearly 200 more years.

The noted science historian Thomas Kuhn recognized this scientific disdain for new ideas when he wrote:

> No part of the aim of normal science is to call forth new sorts of phenomena, indeed those that do not fit the box (containing the prevailing paradigm) are often not seen at all. Nor do the scientists normally aim to invent new theories and they are often intolerant of those invented by others [17].

Our 1993 paper mentioned the influence several previous researchers had on our thinking when we proposed the Alzheimer's vascular hypothesis. Among these were **William Tuke, Ove Hassler, Rachel Ravens, Victor Fischer, and Arnold Scheibel**, who had each noted a variety of microvascular deformities such as kinking, looping, and tortuosity in the brains of deceased psychiatric and demented patients. These investigators were savvy enough to know that such microvessel deformities would make blood flow hemodynamically unstable in the brains of these patients when they were alive but in reading their papers, **I was astonished that not one of these highly competent brain researchers had ever suggested that these microvascular distortions could affect cerebral metabolism in such a way as to induce the cognitive and pathological changes seen in Alzheimer's or dementia. They attributed the microvascular distortions observed to other pathological causes, either tissue atrophy, blood-brain barrier abnormalities, neuronal loss, or neurodegenerative damage. Despite this glaring omission, their work formed the inspiration that led us to the obvious conclusion that twisted microvessels in the brain were partly responsible for poor blood flow and for the inescapable consequences of Alzheimer's. But, there were obviously other factors besides brain capillary distortions involved in the development of cognitive decline and Alzheimer's. What could these be?**

Thinking hard about this missing link to brain hypoperfusion, it became apparent that an association between medical conditions that lower cerebral blood flow and Alzheimer's must exist. Based on our previous evidence and how I envisioned the pathogenesis of Alzheimer's, I decided to take an educated guess and predict in my 1994 [18] **and 1997** [19] **papers that disorders, such as atherosclerosis, type 2 diabetes , brain ischemia, heart disease, hypertension, hypotension, a prior head injury as well as aging, were major vascular risk factors to Alzheimer's disease. These vascular risk factors were later confirmed from epidemiological studies as disorders that lowered blood flow to the brain and which appeared to be associated with cognitive impairment in the elderly** [20].

Our proposal that Alzheimer's was a vascular disorder with neurodegenerative consequences also took an upturn from a collaborative study between us and **Professor Paul Luiten** of Groningen University in Holland. **Professor Luiten's study revealed that 12 months of brain hypoperfusion induced by our carotid artery ligation model in old rats led to damaged cerebral microvessels and progressive memory loss** [21]. This finding suggested that chronic brain hypoperfusion was capable of damaging microvessels and disturb their flow of blood and, this in turn, was responsible for the observed memory deficits. One wonders if the rat evidence had been available to the microvascular morphologists, whether their conclusions might have been different on the role of brain hypoperfusion in the development of Alzheimer's.

The field of neuropathology in the 1970s and 1980s was also influenced by the belief that neurons controlled brain blood flow. That belief seemed to explain that when facing death from trauma or disease, neurons no longer need energy nutrients carried by the microvessels which would then atrophy and disappear from the brain tissue affected. **There was a fallacy in this thinking**. It implied that dying brain neurons would commit suicide by turning off blood flow out of disdain for life and calmly accept death. First, in our judgment, this was an error in anti-Darwinian thinking for which no convincing evidence existed. Second, vascular risk factors specifically associated with Alzheimer's were not considered key contributors to the development of Alzheimer's until we suggested this possibility in our 1994 and 1997 papers, and even then, it was mostly ignored [18, 19].

The idea that microvessels could carry less blood to the brain from cardiac anomalies derived from cardiovascular disease or atherosclerosis was not accepted at the time by most investigators, and even now, many still believe that vascular factors have no influence on Alzheimer's onset. This denial that Alzheimer's appears to be caused by vascular pathology affecting cognitive function has marginalized a number of young researchers seeking funding from government and private foundation grants. More importantly, it is a stumbling block in our ability to move forward and offer potential Alzheimer's patients some hope in circumventing a dreaded fate.

The time had come to clear the air in 1997 and organize an international conference on vascular issues that were deemed relevant to the evolution of Alzheimer's. I needed to know that there were at least 100 investigators in the planet who thought brain blood flow was an important issue in neurodegeneration, cognitive function, and aging.

This topic had never been discussed at any meeting before. I managed to convince the prestigious New York Academy of Sciences to host the meeting and allow me to chair the conference. To make sure the meeting was well attended, I asked **Dr. Vladimir Hachinski**, a well-known neurologist from Canada to be my cochairman. "You mean you want to do all the work while I take all the credit?" asked Hachinski half-jokingly, "you bet," I replied. As it turned out, Hachinski did his part of the work very well and the meeting was a huge success. Sixty presentations on the subject of "Cerebrovascular Pathology in Alzheimer's Disease" [22] were given by a panel of Alzheimer's disease experts from 18 different countries to an audience of more than 600 attendees. It appeared that some vascular issues related to Alzheimer's were here to stay.

Nonetheless, it took several more years for the realization to slowly sink in that cerebral perfusion deficits had a pivotal impact on Alzheimer's pathology and cognitive function. The chief opposing force that kept the vascular hypothesis of Alzheimer's disease out of mainstream research was the Abeta hypothesis which now was at the height of its fishing expedition in search for a magic drug that could turn irreversible brain disintegration into a moneymaker. Several pharmaceuticals were dumping hundreds of millions of dollars into exploiting the notion that an anti-Abeta drug could cure or highly benefit patients whose brain had already been destroyed by an unrelenting neurodegenerative process left over from years of Alzheimer's fury. **It both amused and befuddled me to see how the suckers eagerly lined up to buy pharmaceutical shares based on medical promises that could not be kept and were clearly dishonest**.

Paradoxically, the repeated clinical failures involving ant-Abeta therapy to treat Alzheimer's pathology have not unhinged its colossal influence among many clinicians who are still persuaded that by getting rid of Abeta in Alzheimer's brains, significant comfort should be seen in the treated patients [23, 24]. This never happened and probably never will. The entire research program structured around anti-Abeta therapy has always hung on dubious facts. It reminds me of a quote attributed to Albert Einstein, "if the facts don't fit the theory, change the facts."

When Neurons Struggle for Survival

The year 2000 may have been the beginning of a turning point for the idea that vascular risk factors were directly involved in the evolution of cognitive decline and Alzheimer's dementia. The same year, the CATCH hypothesis was published where I stated that advanced aging in the presence of vascular risk factors can converge to create a **critically attained threshold of cerebral hypoperfusion (CATCH)** that triggers hemodynamic disturbances in the brain, and by so doing, impairs optimal delivery of energy substrates needed for normal nerve cell function; this outcome generates a chain of events leading to a progressive cascade of neurometabolic, cognitive, and tissue pathology that characterizes Alzheimer's disease [25]. Irreversible neuronal damage or death occurs when CATCH is reached due to a severe deficit in energy supply and utilization.

The CATCH hypothesis did not replace the vascular hypothesis, but rather, like any useful concept, it extended and clarified how brain hypoperfusion in an aging person can worsen over time [25] (Fig. 14.2). It will be recalled from Chap. 1, (see Cerebral hypoperfusion) that as a person ages, blood flow to the brain **normally** declines gradually and is noted to drop about 20 % from age 22 to 62 [26]. This may not be a problem as long as vascular risk factors (Figs. 12.1–12.3) to Alzheimer's are not present. If that happens, an additional burden is added to the age-related decline of cerebral blood flow. When cerebral blood flow in the aging individual is reduced to a critical threshold (CATCH), brain cells, especially those that regulate cognitive activities in specific brain regions (because they work the hardest and need more blood flow), may not be able to sustain normal function. It is like adding a heavy pack of wood to a burro's back, then adding some more wood until the burro collapses.

The process involved in CATCH neatly accounts for how aging, cerebral blood flow, brain cell activity, neurotransmission, and cognitive function **interact** with each other to predispose either good or bad aging. One way to sustain good aging in

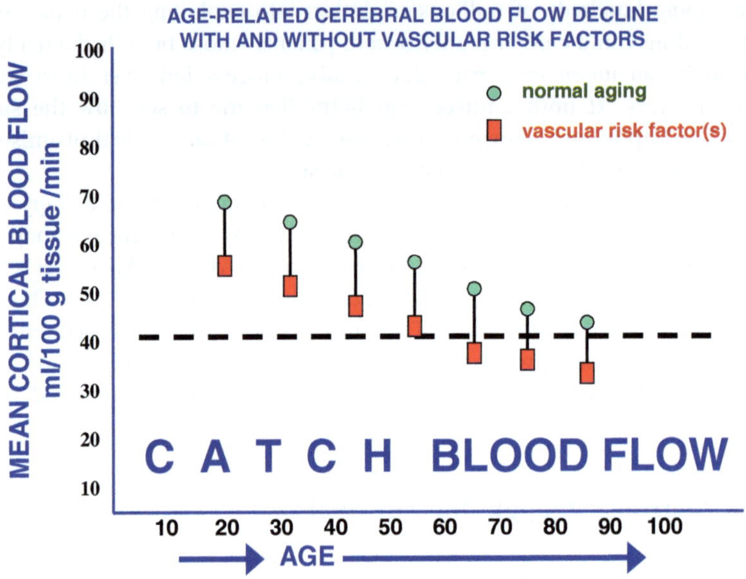

Fig. 14.2 Cerebral blood flow (CBF) normally declines during aging at a rate of approximately 0.05 % per year (*green circles*). From age 20–60, there may be a normal 20 % drop of CBF. Vascular risk factors (VRF) to Alzheimer's (see Figs. 12.1–12.3) will further reduce CBF depending on the duration, type, and severity of the VRF. A critically attained threshold of cerebral hypoperfusion (CATCH) generated by VRF can hypothetically accelerate and worsen CBF (*red squares*) to a point where delivery of CBF to brain cells is markedly reduced. CATCH may signal a point of no return in brain cell survival, a process that can ostensibly lead to severe cognitive impairment and progressive neurodegeneration

the presence of Alzheimer's vascular risk factors is to treat or manage such factors. For example, high blood pressure, type 2 diabetes, and atrial fibrillation can be treated while smoking, not exercising, and obesity can be managed through lifestyle changes. When treating or managing Alzheimer's risk factors becomes impractical or does not work, bad aging can result with dire consequences. The physiopathological sequence involving bad aging and Alzheimer's conversion is represented in Fig. 14.3.

The key to understanding how Alzheimer's develops is to understand how cerebral blood flow behaves during advanced aging. To this end, it is important to know some basic rules about the circulatory system. Circulation in the body is a transport system that keeps us alive. Blood flow transports and delivers vital nutrients, metabolites, clotting factors, and antibodies to all organs of the body. Delivery of oxygen and glucose provides the molecules necessary to create ATP, the chemical that fuels all mammalian cells. The first requirement in a transport system that pushes bulk fluid through a series of pipes is a reliable pump. The human heart does this remarkably well, rhythmically pumping blood into the circulation 100,000 times per day and 35 million times per year.

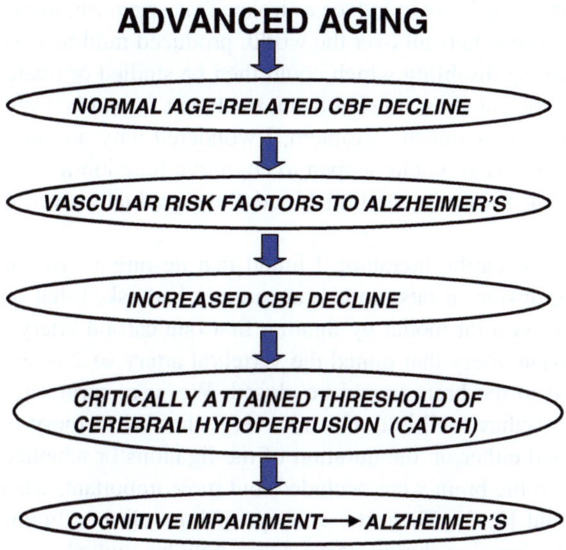

Fig. 14.3 Hypothetical model of how normal aging may advance to cognitive impairment and Alzheimer's. The model is partly based on evidence-based findings linking advanced aging plus the acquisition of vascular risk factors that can result in further cerebral blood flow (CBF) decline. The triad composed of advanced aging, vascular risk factors to Alzheimer's, and declining CBF will provoke CATCH, a hemodynamic alteration which can damage brain cells and lead to cognitive loss and Alzheimer's disease [25]. (see also Fig. 14.2)

The heart can adjust its blood flow output when necessary and does not need the brain to tell it what to do. It can keep on beating without the brain's help, as evidenced by brain death or high cervical spinal cord transection that isolates nerve pathways from the brain to the heart. The brain's ventrolateral medulla can and does influence the heart rate by releasing neurotransmitters into the blood stream to slow or speed up the number of heart beats, for example, in *fright, flight, flight reactions*. Otherwise, the heart has its own electrical system composed of tiny cells in the sinoatrial node located at the top of the right atrium that sends electrical signals to the lower atrioventricular node to initiate the contraction of the entire heart in unison and create a heartbeat.

When the heart stops pumping, circulation shuts down the delivery of nutrients necessary for brain cell survival and death of these cells follows within minutes. **But, what happens when the circulation only slows down as a result of disease or of normal aging?** Curiously, although tens of thousands of scientific articles have been published in the last 100 years about the physiology of blood flow, very few papers prior to our vascular hypothesis of Alzheimer's proposal in 1993 [16] have focused on this question.

The way this came about is worth mentioning. My research for many years, prior to my involvement in Alzheimer's, concentrated on problems related to stroke which could be experimentally induced in animals by permanently occluding a major artery within the brain to induce a stroke. This approach, used in thousands of experiments by researchers all over the world, produced mild-to-major neurological deficits and learning disability which could then be studied or treated. Despite tens of thousands of such studies, no significant treatment for stroke ever came out of it. One day, in thinking about this problem, I wondered why no one, it seemed, had asked the question, what would happen to cognitive function if instead of a sudden stroke, one where to lower cerebral blood flow gradually over many weeks or months.

Searching the scientific literature, I found that despite a vast variety of articles using carotid occlusion in rats, no one apparently had asked that question. So, we decided to try it on a rat model by ligating first one carotid artery, then two, then added a subclavian artery that joined the vertebral artery so 2 or 3 of the 4 vessels supplying blood to the brain would be closed. We were surprised to see that rats tolerated the procedure well but, after recovery, showed memory problems whose severity depended either on the duration of the ligations or whether 2 or 3 arteries supplying flow to the brain were occluded, but most important, whether the animal was young or old [2, 3]. Thus, the concept of how chronic brain hypoperfusion disturbs memory function during aging could now be studied.

The role of long-term cerebral hypoperfusion is now a topic of increasing research interest, especially as it involves aging, cognitive function, neurodegeneration, and development of Alzheimer's. High-resolution brain scans performed in the last 10 years have documented the role of brain hypoperfusion at different stages of cognitive decline and Alzheimer's. For some time now, it has been known that cerebral blood flow is reduced during Alzheimer's.

Subscribers of the Abeta hypothesis ascribed this to dying or dead neurons turning off local blood flow because they no longer had any need for it. This seemed counterintuitive because it implied that nerve cells chose not to fight damage like other cells in the body in the face of unrelenting disease. If this were the case, all the regulatory mechanisms that keep nerve cells alive when blood pressure is altered, or when cerebral perfusion is deficient or when stress or injury strikes, would be pointless.

A Summary of the Evidence

An overview of our blooming scientific hypothesis that took a life of its own began with some experiments on young and aged rats subjected to reduced cerebral blood flow. These basic studies led us to the following deductions:

A. Reducing cerebral blood flow in aged but not young rat brain leads to memory impairment;
B. When cerebral blood flow reduction in aged rats is maintained for 6–12 months (equivalent to 16–33 human years), brain changes develop including capillary damage, ATP cell energy decline, impaired protein synthesis, death of neurons in cognitive-related regions, and brain atrophy;

C. Structural brain changes and cognitive loss that appear in aged rats with reduced cerebral blood flow mimic human Alzheimer's;
D. Chronic reduction of cerebral blood flow in aging humans appear to trigger cognitive changes that can lead to the development of Alzheimer's disease; this observation engenders the vascular hypothesis of Alzheimer's;
E. Conditions (vascular risk factors) that lower brain blood flow during aging directly contribute to the development of Alzheimer's;
F. To delay or prevent Alzheimer's, vascular risk factors must be managed or treated and ideally, interventions that can increase cerebral blood flow in a sustained and safe fashion could help reverse or prevent cognitive impairment.

The vascular hypothesis of Alzheimer's disease seems to have survived the unpredictable three stages of a scientific discovery. According to the nineteenth-century naturalist Alexander von Humboldt, scientists view such discoveries by first ignoring it, then denying its usefulness, and finally, crediting someone else.

Since the usefulness of a medical hypothesis depends mainly on its pragmatic qualities of improving something or someone, it remains to be seen whether the published observations relevant to the vascular hypothesis will advance our understanding of how dementia evolves and ways to clinically prevent it (Fig. 14.3).

If the **vascular hypothesis of Alzheimer's disease is proven accurate as the cause of Alzheimer's disease, its usefulness may extend to explaining the pathogenesis of other neurodegenerative disorders, for example, frontotemporal dementia, Parkinson's disease, and Lewy body dementia.** There is already a pathogenic overlap between vascular dementia and Alzheimer's even though the former is known to trigger cognitive deterioration by a stroke or many ministrokes, while the latter seems to be generated by chronic brain hypoperfusion. Many patients present with evidence of brain infarcts in the gray or white matter together with abundant deposition of plaques and tangles, a condition called "mixed dementia."

The dementias named above have many common symptoms including cognitive difficulties and although their etiologies are far from clear, it is reasonable to suspect that they may also share the chronic brain hypoperfusion that is typically found in patients who convert from mild cognitive dysfunction to Alzheimer's. **If this is found to be true, it would open the door to a new scientific field that one could call "vasculocognopathy," that is, the investigation of how abnormal blood flow can affect cognitive function,** a term we coined in 2004 [27].

To summarize, when all the evidence presented here is carefully analyzed (and this book only contains a fraction of that evidence), the pathological fingerprint in the evolution of Alzheimer's disease appears to be the sum of two factors: brain hypoperfusion and advanced aging. Since mild brain hypoperfusion is part of advanced normal aging, the pathological state must be reached by the *worsening* of normal brain hypoperfusion and the deliverer must be a substrate that is known to persistently lower brain blood flow, for example, vascular risk factors. This assumption satisfies the concept that Alzheimer's needs a pivotal trigger to initiate dementia and a favorable substrate to maintain it.

My scientific and medical training as well as my experience in the laboratory have taught me the hard lesson of being extremely careful when interpreting basic animal data. I have tried in my publications to avoid the trap of hasty assumptions. In the case of the vascular hypothesis of Alzheimer's disease, I can state with some confidence that I have never been so certain of something in all my life. No hypothesis is ever perfect and the vascular hypothesis to explain the evolution of Alzheimer's is no exception, but, I do not see a lethal piece of evidence that kills the concept in its tracks. I sincerely welcome any contradictory evidence that disproves the fallacy of this interpretation.

I don't doubt that many cynics will default to the prevailing knee-jerk dogma until that time when failure to act is no longer fashionable, profitable, or reasonable. It is predictable that there will be many undiscovered sidesteps to the final solution of the Alzheimer's puzzle, but, that is, what makes research so fascinating.

References

1. Marshall RS. Effects of altered cerebral hemodynamics on cognitive function. J Alzheimers Dis. 2012;32(3):633–42.
2. de la Torre JC, Fortin T, Park GA, Saunders JK, Kozlowski P, Butler K, de Socarraz H, Pappas B, Richard M. Aged but not young rats develop metabolic, memory deficits after chronic brain ischaemia. Neurol Res. 1992;14(2 Suppl):177–80.
3. de la Torre JC, Fortin T, Park GA, Butler KS, Kozlowski P, Pappas BA, de Socarraz H, Saunders JK, Richard MT. Chronic cerebrovascular insufficiency induces dementia-like deficits in aged rats. Brain Res. 1992;582(2):186–95.
4. de la Torre JC, Fortin T. A chronic physiological rat model of dementia. Behav Brain Res. 1994;63(1):35–40.
5. Henderson B, de la Torre JC. Reversal of chronic ischemia in the adult rat: common carotid anastomosis and improvement in memory dysfunction. Soc Neurosci. 1999;25:55.
6. Gustafson L, Hagberg B. Emotional behaviour, personality changes and cognitive reduction in presenile dementia: related to regional cerebral blood flow. Acta Psychiatr Scand Suppl. 1975;257:37–71.
7. Hedlund S, Koheler V, Nylin G, Olsson R, Regnstroem O, Rothstroem E, Astroem KE. Cerebral blood flow in dementia. Acta Psychiatr Scand. 1964;40:77–106.
8. Hagberg B, Ingvar DH. Intellectual impairment in presenile dementia related to regional cerebral blood flow. Act Nerv Super (Praha). 1977;19(Suppl 2):350–1.
9. Davis DG, Schmitt FA, Wekstein DR, Markesbery WR. Alzheimer neuropathologic alterations in aged cognitively normal subjects. J Neuropathol Exp Neurol. 1999;58:376–88.
10. Giannakopoulos P, Herrmann FR, Bussiere T, Bouras C, Kövari E, Perl DP, Morrison JH, Gold G, Hof PR. Tangle and neuron numbers, but not amyloid load, predict cognitive status in Alzheimer's disease. Neurology. 2003;60:1495–500.
11. Price JL, McKeel DW Jr, Buckles VD, Roe CM, Xiong C, Grundman M, Hansen LA, Petersen RC, Parisi JE, Dickson DW, Smith CD, Davis DG, Schmitt FA, Markesbery WR, Kaye J, Kurlan R, Hulette C, Kurland BF, Higdon R, Kukull W, Morris JC. Neuropathology of nondemented aging: presumptive evidence for preclinical Alzheimer disease. Neurobiol Aging. 2009;30:1026–36.

12. Shimada H, Ataka S, Tomiyama T, Takechi H, Mori H, Miki T. Clinical course of patients with familial early-onset Alzheimer's disease potentially lacking senile plaques bearing the E693Δ mutation in amyloid precursor protein. Dement Geriatr Cogn Disord. 2011;32(1):45–54.
13. Verclytte S, Lopes R, Lenfant P, Rollin A, Semah F, Leclerc X, Pasquier F, Delmaire C. Cerebral hypoperfusion and hypometabolism detected by arterial spin labeling MRI and FDG-PET in early-onset Alzheimer's disease. J Neuroimaging. 2015. (In press).
14. de la Torre JC. Does brain microvessel pathology provoke Alzheimer's disease? Soc Neurosci Abstr. 1992;18:564.
15. Fung YC. Biomechanics: circulation. New York: Springer; 1984.
16. de la Torre JC, Mussivand T. Can disturbed brain microcirculation cause Alzheimer's disease? Neurol Res. 1993;15:146–53.
17. Kuhn TS. The structure of scientific revolutions. 3rd ed. Chicago, Illinois: The University of Chicago Press; 1996, p. 24.
18. de la Torre JC. Impaired brain microcirculation may trigger Alzheimer's disease. Neurosci Biobehav Rev. 1994;18(3):397–401.
19. de la Torre JC. Hemodynamic consequences of deformed microvessels in the brain in Alzheimer's disease. Ann NY Acad Sci. 1997;26(826):75–91.
20. Breteler MM. Vascular risk factors for Alzheimer's disease: an epidemiologic perspective. Neurobiol Aging. 2000;21(2):153–60.
21. de Jong GI, Farkas E, Plass J, Keijser JN, de la Torre JC, Luiten PGM. Cerebral hypoperfusion yields capillary damage in hippocampus CA1 that correlates to spatial memory impairment. Neuroscience. 1999;91:203–10.
22. de la Torre JC, Hachinski, V. (eds.) Cerebrovascular pathology in Alzheimer's Disease. Ann NY Acad Sci. 1997;826:1–523.
23. Salloway S, Sperling R, Fox NC, et al. Two phase 3 trials of bapineuzumab in mild-to-moderate Alzheimer's disease. N Engl J Med. 2014;370:322–33.
24. Doody RS, Thomas RG, Farlow M, et al. Phase 3 trials of solanezumab for mild-to-moderate Alzheimer's disease. N Engl J Med. 2014;370:311–21.
25. de la Torre JC. Critically attained threshold of cerebral hypoperfusion: the CATCH hypothesis of Alzheimer's pathogenesis. Neurobiol Aging. 2000;21(2):331–42.
26. Leenders KL, Perani D, Lammertsma AA, Heather JD, Buckingham P, Healy MJR, Gibbs JM, Wise RJS, Hatazawa J, Herold S, Beaney RP, Brooks DJ, Spinks T, Rhodes C, Frackowiak RS, Jones T. Cerebral blood flow, blood volume and oxygen utilization. Brain. 1990;113:24–47.
27. de la Torre JC. Alzheimer's disease is a vasocognopathy: a new term to describe its nature. Neurol Res. 2004;26(5):517–24.

Chapter 15
Clinical Tools to Detect and Predict Individuals at Risk of Alzheimer's

Cerebrovascular Evaluation with Neuroimaging

The first decade of the twenty-first century has seen a growing recognition that low blood flow to the brain (cerebral hypoperfusion) and abnormal hemodynamics of the aging brain are directly related to the development of cognitive deficits as precursors of Alzheimer's [1]. It is now appreciated that vascular risk factors for Alzheimer's potentiate the formation of Abeta in the brain [2, 3]. **This finding offers proof of concept that Abeta deposition in the brain is not the cause of Alzheimer's but a pathological product, ostensibly formed from protein aberrations promoted by chronic brain hypoperfusion and the neuronal energy crisis that follows. Verification of this conclusion would explain the paradoxical appearance of heavy Abeta accumulation in both the brains of cognitively intact and cognitively impaired aging individuals.**

Two things can be said about Alzheimer's disease with some degree of confidence. First, there is now a consensus of opinion that the underlying pathology involved in the development of Alzheimer's begins in the normal individual 10 or 20 years before dyscognitive symptoms are expressed. Most vascular risk factors for Alzheimer's appear at middle age or even before and continue into advanced aging. This clinical setting is consistent with the progressive brain hypoperfusion associated with vascular risk factors during aging. Second, since Alzheimer's is also considered by many investigators to be a heterogenous disorder (composed of dissimilar elements), vascular risk factors with their discordant etiologies and divergent disease courses appear to be auspicious contributors to the evolution of Alzheimer's.

It is important to note that the Alzheimer's risk factors as listed in Figs. 12.1–12.3 **have one thing in common, they all reduce cerebral blood flow**. The probability that several dozen of these risk factors are capable of reducing cerebral blood flow out of coincidence is unlikely since they do not otherwise share another common pathological factor that unites them.

J.C. de la Torre, *Alzheimer's Turning Point*,
DOI 10.1007/978-3-319-34057-9_15

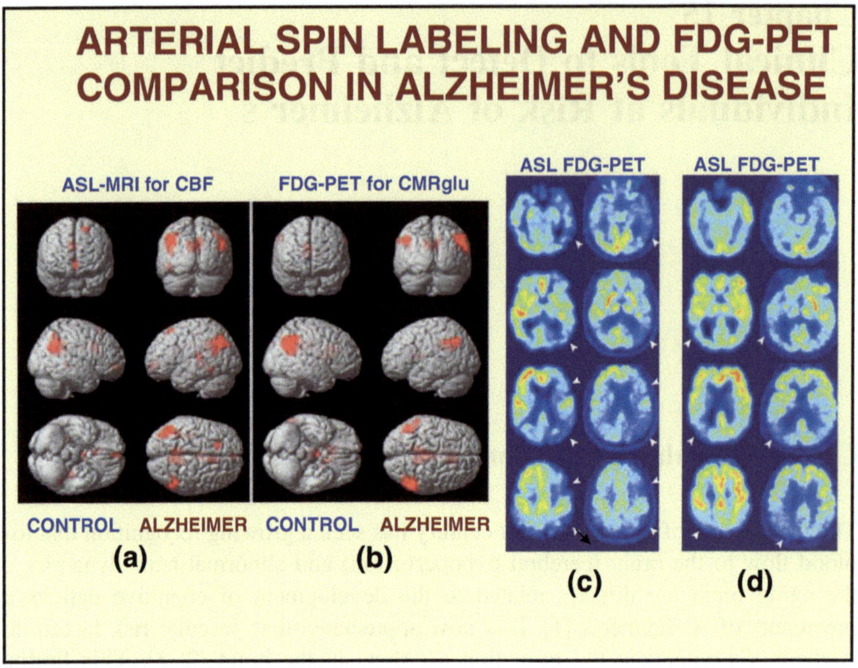

Fig. 15.1 Topographical brain areas of cerebral hypoperfusion (**a**) and low metabolism (**b**) in early Alzheimer's disease (marked in *red*), using arterial spin labeling MRI (ASL) measuring cereberal blood flow (CBF) and fluorodeoxyglucose positron emission tomography (FDG-PET) measuring cerebral metabolic rate of glucose (CMRglu) as compared to cognitively intact controls. Note greater areas of hypoperfusion and hypometabolism relative to controls in similar brain regions. Intracerebral (**c**, **d**) regions of hypoperfusion and hypometabolism (small *white arrowheads*) between ASL and FDG-PET in an Alzheimer's brain. Note parallel distribution between low brain blood flow (**c**) and reduced glucose uptake (**d**), a marker of hypometabolism. **a**, **b** adapted from Chen Y, et al. Neurology 2011:77:1077–1985. **c**, **d** adapted from Musiek ES, et al. Alzheimer's Dement 2012;6:51–59

If cerebral blood flow reduction from vascular risk factors equates to a close association with Alzheimer's onset, the logical inference that follows could be quite meaningful in clinical terms. For example, if this association is valid, it would be wise to have the necessary diagnostic tools that can detect the presence of such risk factors at the earliest stage preceding cognitive impairment.

There is now considerable evidence that cognitive deterioration and eventual development of Alzheimer's can be predicted using neuroimaging tools that have the ability to accurately measure cerebral blood flow in the patient at risk of dementia. Brain neuroimaging using the marker (18F)fluorodeoxyglucose positron emission tomography (FDG-PET) has shown reduced glucose uptake in the regions of the brain later targeted by neurodegeneration and pathologic formation of senile plaques and neurofibrillary tangles [4, 5]. These hypometabolic brain areas include a network of parietal, posterior cingulate, temporal, and frontal regions (Fig. 15.1).

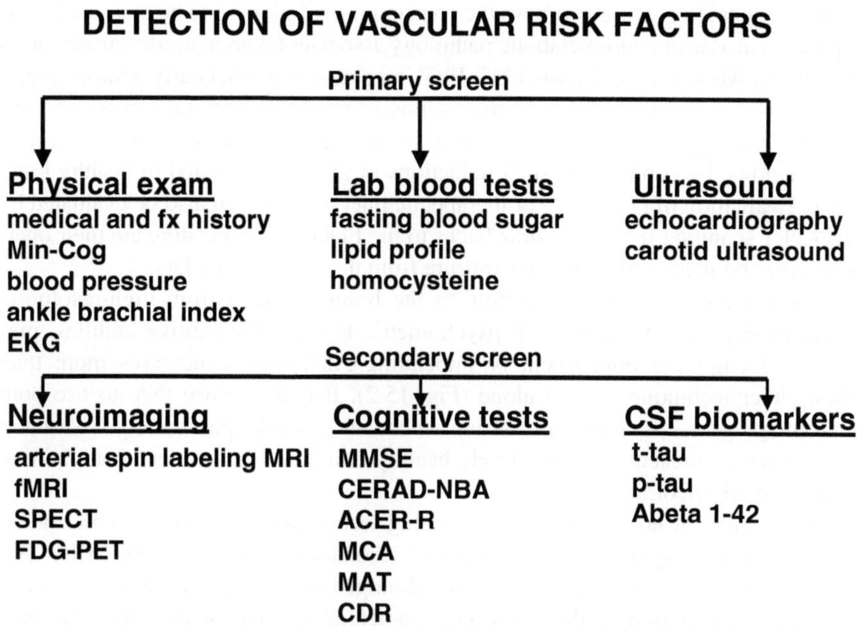

Fig. 15.2 Algorhithmic primary screen examination and testing of a patient arriving at the heart brain clinic. A physical examination includes a personal and family (fx) history of cardiac or mental disorders. A Mini-Cog test can be done in 3 min to screen older adults for cognitive impairment. Measures of blood pressure, peripheral artery disease (ankle–brachial index), and electrocardiogram (EKG) determine the presence of cardiac-related vascular risk factors. After routine laboratory blood work, cardiac function is examined with an echocardiogram and carotid artery patency is tested with Doppler ultrasound. Cerebral imaging, cognitive tests, and cerebrospinal fluid markers can be performed if primary screen detects cardiac, peripheral artery abnormalities, or cognitive dysfunction

Reduced glucose uptake in these FDG-PET studies indicates the presence of cerebral hypoperfusion that results in lower metabolic function typically in brain regions such as the posterior cingulate gyrus, the precuneus, the prefrontal and parietal cortices, and the temporoparietal cortex (Fig. 15.1).

99m Tc-hexamethylpropyleneamineoxime single-photon emission computed tomography (SPECT) has also been used as an early marker of Alzheimer's. SPECT has been used to identify individuals with mild cognitive impairment who show a preclinical risk to Alzheimer's conversion [6]. SPECT scanning was able to provide useful diagnostic information that could support the clinical examination and confirm, to a degree, the risk to Alzheimer's. However, SPECT is now considered less reliable than MRI or PET scanning because of its lower resolution and higher variability of cerebral blood flow measurements.

PET imaging can involve using various metabolic markers including the radioactive analogue of glucose, 18F-labeled fluorodeoxyglucose (FDG). The FDG-PET technique involves injecting the radiotracer FDG and placing the patient in a PET scanner to image the brain. Since FDG is a glucose analogue with physiological properties nearly identical to glucose, it allows the consumption of glucose to be measured. This approach can detect neurometabolic pathology associated with neurodegeneration, as it occurs in Alzheimer's disease. FDG-PET imaging can predict early neurodegeneration of specific cognitive-related brain regions even before structural damage of those regions occur [7]. It does this by measuring whether glucose metabolism is reduced in those regions; by contrast, no such reductions in glucose are noted in healthy brain tissue. Evaluation of preventive or therapeutic interventions can also be evaluated by FDG-PET scanning [8]. Major drawbacks to the FDG-PET procedure are their operating cost and increased radiation exposure from the radiotracer FDG.

When findings of low blood flow to the brain obtained from high-resolution neuroimaging are combined with psychometric testing of cognitive abilities, prediction of cognitive impairment or impending Alzheimer's increases more than when either technique is used alone (Fig. 15.2). It makes sense then to use both neuroimaging and psychometric testing whenever possible, particularly if vascular risk factors are present in a cognitively healthy individual or someone complaining of memory shortcomings.

One recent and noninvasive neuroimaging technique to measure cerebral blood flow is arterial spin labeling magnetic resonance imaging (ASL-MRI) [9]. ASL-MRI is considered by many neuroradiologists to be an indispensable tool in the clinical evaluation of the brain and a potential method of detecting and predicting cognitive decline and conversion from mild cognitive impairment to Alzheimer's onset. ASL-MRI has been reported to predict clinical function and cognitive decline in people at risk of Alzheimer's by detecting the presence of brain hypoperfusion in various brain regions as discussed above [10] (Fig. 15.1).

A dramatic prediction in the development of cognitive difficulties using ASL-MRI was made recently by a team from the University of Geneva led by **Dr. Sven Haller**. Healthy asymptomatic elderly individuals were evaluated for cerebral blood flow deficits. It was found that reduced brain blood flow in the posterior cingulate cortex of cognitively intact patients led to subsequent cognitive deterioration 18 months later [11]. If this study is confirmed, the finding could be one of the most important contributions in the evaluation of asymptomatic older patients and their potential risk of developing Alzheimer's dementia. It would also allow a complete neurological and neurocognitive evaluation of patients, as discussed in Chap. 16, to apply a clinical plan in determining the presence of vascular risk factors that could be associated with the reduced cerebral blood flow and administer, at a very early stage, an appropriate treatment or management that can reduce such risk factors.

The advantages of ASL-MRI over PET or SPECT are that there is no exposure to radioactive tracers or ionizing radiation since it only labels endogenous arterial blood water which is cleared from the system within seconds [12]. ASL-MRI can be performed in most MRI scanners and takes less than 15 min to obtain results. Since it is important to accurately quantify whether brain hypoperfusion is present in the patient being tested, ASL-MRI offers a reliable, fast, and sensitive method for determining the potential risk to cognitive decline and eventual dementia.

Needless to say, the use of ASL-MRI opens a window of opportunity for testing potential therapy that could prevent further progress of the cerebral blood flow decline. Functional MRI (fMRI) is a procedure similar to MRI, but in addition to measuring brain blood flow, it will also indirectly measure neural activity globally or regionally. This approach, called blood oxygen-level dependent, or BOLD, can localize brain activation in a region several millimeters long by measuring changes in the proportion of oxygenated (oxygen-rich) and deoxygenated (oxygen-poor) blood in the capillaries that supply nerve cells. One can then infer any changes in neural activity due to cognitive dysfunction, age-related hemodynamic abnormalities, or dementia. For example, when a neural event such as recalling a memory occurs in the brain, microvascular blood flow feeding the region of the neurons responsible for the memory activity increases, while deoxygenated blood in the same microvessels decreases. Thus, the BOLD signal increases. This phenomenon is known as **neurovascular coupling** where the brain blood supply meets the energy demand of the interacting neurons, glia, and vascular cells. This phenomenon is discussed in Chap. 10.

Neurovascular coupling and the law of supply and demand form the basis of the vascular hypothesis of Alzheimer's that we proposed in our 1993 paper which stipulated that a neuronal energy crisis consisting of low cerebral blood flow to the brain not commensurate with energy demand is the primary cause of progressive cognitive impairment and Alzheimer's [13].

As the neuroimaging technology improves, its value in detecting and predicting cognitive deterioration in the elderly that can interfere with daily activities becomes a crucial target for interventions. Such interventions can be created to delay or prevent serious cognitive meltdown. The evidence gathered so far from these neuroimaging studies show that cerebral blood flow may be an excellent marker to predict impending cognitive decline and conversion to Alzheimer's. In the past few years, a barrage of well-documented neuroimaging studies has firmly established the clinical value of blood flow measurements of the brain with respect to prognosing cognitive decline. These brain scan studies open a window of opportunity for interventions that can prevent irreversible damaging changes of the brain and potentially improve the patient's clinical outcome.

The process of detecting patients at risk of Alzheimer's prior to any cognitive impairment is dependent on knowing the cause of this dementia. **If, as we suspect and have proposed, vascular risk factors during advanced aging are what pushes cerebral hypoperfusion to create cognitive impairment and eventually Alzheimer's disease** (Fig. 14.3), **then diagnostic tools that can detect such risk factors need to be employed** (Fig. 15.2). Looking at Fig. 15.2, a primary screen using equipment found in most medical offices offers a first-step impression of whether the patient would benefit from a more detailed secondary screen. Preliminarily, a physical examination would first obtain personal and family history of mental disease, together with blood pressure measurement, an electrocardiogram, and any sign of peripheral artery disease using the ankle–brachial index.

A physical examination can determine heart, lung, and neurological abnormalities that may impact cognitive function. Patient history can reveal lifestyle risks to normal mental health such as family history of dementia, alcoholism, prior traumatic brain injury, poor nutrition, mental and physical inactivity, current and past illnesses, and patient medications. Routine blood tests can uncover possible risks to cognitive impairment in the elderly such as diabetes type 2, thyroid abnormalities, anemia, vitamin deficiencies, and serum lipid profiles that can lead to atherosclerosis, heart disease, and stroke. **Special blood tests for $APOE_4$, clusterin, $Abeta_{42}$, and homocysteine can be considered if cognitive impairment is detected**. Patient history or chief complaint determines the sequence of brain and heart testing that needs to be done. Positive and negative findings from both heart and brain clinics' testings are reviewed and interpreted, and a medical plan of action or observation period with follow-up is chosen. Periodic follow-up visits by patients determine any further changes in the medical plan of action or in continued observation.

Cardiovascular Evaluation

Noninvasive ultrasound of the heart using echocardiography can reveal how blood flows through the chambers of the heart and heart valves and how effective the heart pumps blood into the aorta. An echocardiogram can also show the thickness of the heart muscles which can put a strain on the heart's pumping action and deliver less blood flow to the brain and other organs than is needed. Indications for a secondary screen (Fig. 15.2) are dependent on the results of the primary screen, the age and status of the patient, and the physician's judgment for further testing.

Another useful cardiac technique that can be used to determine the stress put on the cardiac muscles is the rate pressure product (RPP). This test measures myocardial oxygen demand and thus is a good measure of the energy consumption of the heart. The test uses the equation RPP = heart rate x systolic blood pressure. Heart rate and systolic blood pressure rise with physical or mental activity so it becomes relatively easy to predict their levels of efficiency using the RPP test. The RPP test is considered one of the best measures of cardiovascular fitness partly because it represents a hemodynamic predictor of the amount of oxygen transported and used in cellular metabolism [14].

Echocardiography is a tool that has been underused in the evaluation of patients at risk of acquiring Alzheimer's. This cost-effective, noninvasive, and powerful diagnostic instrument that is commonly used in the global evaluation of cardiac structure and function cannot be understated. It is elementary that a desktop tool that can determine cardiac output and cardiac index would be invaluable in assessing whether brain hypoperfusion is cardiac-dependent or not (Fig. 15.3). But echocardiography does much more than that.

There are 4 types of echocardiographic approaches generally used clinically: (a) M-mode, (b) 2-dimensional (or B-mode), (c) 3-dimensional, and (d) Doppler echocardiography. M-mode echocardiography is useful in measuring the left and right ventricular function, valvular abnormalities, and cardiac hypertrophy [15]. Two-dimensional echocardiography images are commonly used to assess the mechanics of left ventricular wall motion, filling chamber dynamics, and atrial contraction. Doppler analysis is clinically very useful because it provides the

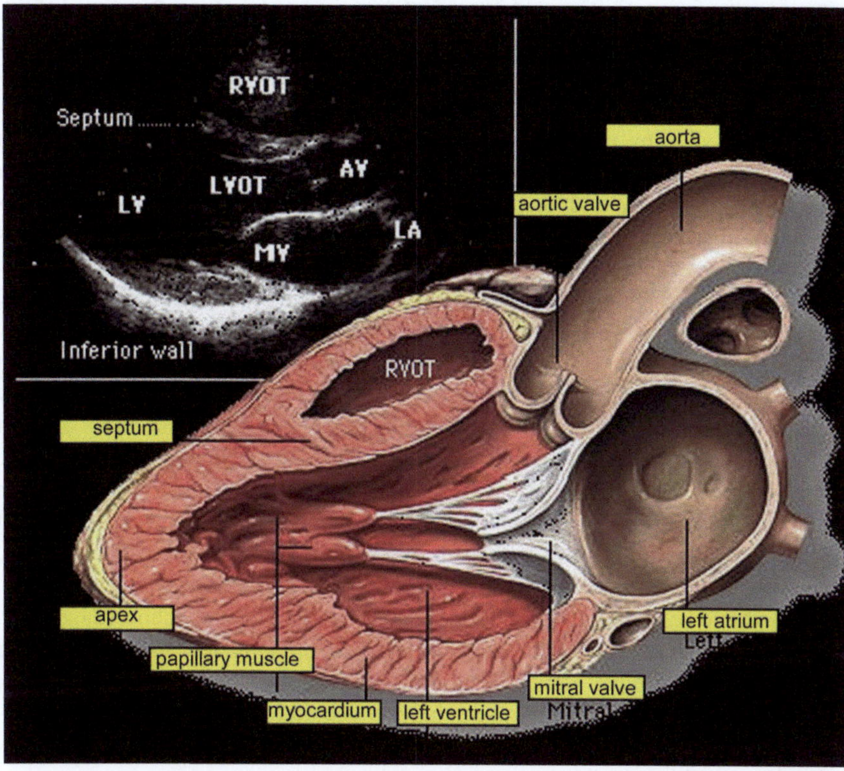

Fig. 15.3 Transthoracic echocardiography showing parasternal long axis view (*inset*) obtained by placing and moving ultrasound probe on surface of chest over heart. This and other cardiac views are useful for detecting and measuring chamber enlargement, valvular disease, pulmonary artery pressure, left ventricular (LV) systolic/diastolic function, ejection fraction, and cardiac output. These cardiac measurements offer noninvasive, cost-effective, and a reliable approach that can identify often reversible or treatable early heart lesions that may contribute to chronic brain hypoperfusion, a common preclinical precursor of Alzheimer's in elderly patients. Key: *AV* atrial valve; *LA* left atrium; *LVOT* left ventricular outlet tract; *LV* left ventricle; *MV* mitral valve; *RVOT* right ventricular outlet tract

examiner with dynamic information on the velocity and direction of blood flow in a region of interest within the normal or pathologic heart (Fig. 15.3). Three-dimensional echocardiography has been shown to have several advantages over 2D echocardiography, especially in determining left ventricular dysfunction and for aortic or mitral valve disease evaluation [16]. Thus, hemodynamic data can be correlated to cardiac structure using 3-dimensional, 2D, M-mode, or combinations of each method, thereby providing more relevant information on the anatomic and functional state of the heart [15, 16]. An important aspect of echocardiography is that most office procedures using the Doppler method can also evaluate carotid artery blood flow and spot any abnormalities that can impede or block brain perfusion.

Echocardiography is also useful in guiding therapy over time. Consequently, if cardiac anomalies that pose a risk to brain perfusion in the elderly individual are found during echocardiography, therapy to correct such anomalies can be monitored periodically and modified if needed.

This is possible by measuring the effect of therapy on multiple parameters of cardiac function, including diastolic filling and ventricular performance, two major determinants of *preload* and, therefore, of cardiac output [17]. These measurements have demonstrated considerable prognostic value in symptomatic and asymptomatic patients with either preserved or abnormal cardiac function [18].

Taken together, echocardiographic measurements can provide a relatively simple, safe, accurate, and cost-effective evaluation of impaired cardiac function that may indicate a potential or an active dangerously reduced level of cerebral perfusion that can lead to mild cognitive impairment and dementia in middle aged or elderly people. More importantly, given the compelling evidence that some forms of heart and carotid artery disease are common findings in elderly patients and that chronic cerebral hypoperfusion secondary to these conditions appears to be a prodromal sign to Alzheimer's and, to a lesser extent, to vascular dementia [19–21], we propose that cognitively healthy individuals with even subjective mild memory recall symptoms, sometimes dismissed as a "senior moment" or benign senescent forgetfulness, should undergo screening using carotid Doppler ultrasound and echocardiography.

Although other cardiac techniques and markers are available to assess cardiac function [22], we have deliberately chosen echocardiography as the initial, cost-effective, noninvasive, safe, and relatively accurate tool that can provide evidence for further workup and prompt intensive therapy in the patient deemed at risk to develop cognitive decline.

Psychometric Testing

There are many reliable and sensitive cognitive function tests that can flag potential and overt cognitive loss, among them the Mini-Mental State Examination (MMSE), Mini-Cog, CERAD Neuropsychology Assessment Battery (CERAD-NBA), Addenbrooke's Cognitive Examination Revised (ACER-R), Montreal Cognitive Assessment (MCA), Memory Alteration Test (MAT), and the Clinical Dementia Rating (CDR) scale (Fig. 14.3). Some cognitive function tests can be done within 15 min or less, and many correlate with cerebrovascular and cardiovascular pathology (Fig. 15.2).

These neuropsychological tests can evaluate and diagnose a variety of cognitive problems or deficits related to memory, learning, intelligence, language processing, attention, psychomotor speed, speed of information, depression, and anxiety. Testing can provide a picture of the patient's strengths and impairments and indicate what treatments or lifestyle changes might reduce or reverse the cause of the mental deficits found. The reliability, sensitivity, and specificity of each test can be enhanced when data from other cognitive tests are available. Moreover, psychometric testing can be a powerful diagnostic tool when combined with neuroimaging measurements of blood flow and metabolism of the brain.

Ocular Screens

Ocular diagnostic techniques are being used more widely in the detection and screening of Alzheimer's using changes that can be measured by ophthalmoscopy of the retinal nerve fiber layer and optic nerve. An excellent review and discussion of these ocular techniques can be found in **Drs. Thomas Lewis and Clement Trempe's** book *The End of Alzheimer's* [23].

Retinal manifestations of preclinical Alzheimer's offer a veritable window of diagnostic potential because one can directly examine a part of the brain noninvasively by instruments which allow peering into the fundus of the eye to visualize the retina and optic nerve. Tools such as optical coherence tomography (OCT) accurately measure neuropathological features within the eye such as retinal vascular abnormalities, and secondary retinal ischemia which have been reported in Alzheimer's dementia. A hot topic of research these days is the study of retinal neurodegeneration, seen through the reduced thickness of the progressive nerve fiber layer (RNFL) and retinal ganglion cells (RGC) in mild cognitive impairment, a transient stage preceding Alzheimer's disease [24]. Thinning of the RNFL in mild cognitive impairment patients implies a loss of retinal neurons and their axons and may be a clue that active or potential cerebral neurodegeneration is present.

Signs of retinal vascular abnormalities using OCT have also been reported to be associated with cognitive decline characterized by reduced executive function and psychomotor speed [25]. This is a key finding in Alzheimer's research because it allows a direct look into a part of the brain (the eye fundus) that may reflect microvascular pathology in the brain tissue itself that cannot be seen noninvasively. The simplicity and cost-effectiveness of ocular diagnostic screens for cognitive dysfunction that can lead to dementia could become a boon in the search of new therapy that can reverse or slow down cerebrovascular disease and brain degeneration.

There is no single test that can consistently predict cognitive decline or Alzheimer's. The diagnostic techniques reviewed above, especially when combined judiciously, can present a very accurate picture of what may be expected within a few years for a brain healthy patient undergoing a battery of cognitive and imaging tests. The technology of Alzheimer's detection no doubt will improve substantially in the next few years but for now, it is wise to apply what we know and have learned about this dementia. **We know, for example that decades before serious cognitive difficulties appear in the aging individual, something in the brain is not being processed normally**. Psychometric testing may detect some cognitive anomalies at a stage where neurological signs are not yet apparent, but this information does not address the possible cause of such anomalies. Moreover, some psychological tests are not as effective as other tests in predicting memory loss due to disease rather than normal aging. Some neuropsychological tests are specific for dementia. For example, Addenbrooke's Cognitive Examination can distinguish frontotemporal dementia from Alzheimer's, but the fact that dementia is already present does not help in planning for a treatment to delay progressive neurodegeneration which in all likelihood has already begun.

Poor blood flow to the brain initiated by vascular risk factors in the presence of age-related lowered cerebral blood flow appears as the most likely activator of cognitive deterioration. This conclusion is supported by substantial animal data discussed in Chap. 10 and by human evidence in older people showing that managing or treating vascular risk factors will lower the incidence of cognitive decline [26]. Treating vascular risk factors such as diabetes, dyslipidemia, hypertension, atherosclerosis, and obesity at any age is medically unassailable, but it becomes urgently required when it is detected at midlife because it could result in primary prevention of Alzheimer's. This conclusion is supported by major community-based studies such as the Rotterdam Study, the Framingham Study, the Cardiovascular Health Study, and the Kungsholmen Project [26].

Where Do We Go from Here?

How will the knowledge gained in these longitudinal studies on vascular risk factors be best applied to prevent cognitive deterioration at midlife? It is not enough to instruct a patient who is obese to cut down the calories or to instruct a couch potato to exercise more. It is said that television's 1970s weekly medical drama Marcus Welby, MD, had an impeccable bedside manner and precise, lifesaving measures because he only had to care for one patient per week. Physicians nowadays see as many patients in their practice as is decently possible, but rarely spend more than 10 min on patients who have been seen before. When I was working in the emergency room in a large, county hospital, I saw and treated not less than 60 patients in a 12 h shift, as did other residents. In some European countries which practice socialized medicine, physicians are generally paid the same salary whether they see 2 patients a day or 200. That gives them the opportunity to spend more time on a patient that clearly needs more attention than 10 min. The result is better medicine. That type of medicine is unlikely to be adopted in the USA where physicians get paid for the volume of patients they see.

The point to be made here is that in order to defer the catastrophic tripling of Alzheimer's prevalence by 2050 and bankrupt Medicare and Medicaid by doling out more than a trillion dollars a year for the care of 14 million new Alzheimer's patients, some key changes in medical practice need to be put forward. **It cannot be overstated that a medical crisis of cataclysmic proportions will swallow healthcare systems in civilized societies unless a plan is applied that will significantly reduce the anticipated Alzheimer's boom**. In Chap. 16, such a plan is described.

References

1. Heinzel S, Metzger FG, Ehlis AC, Korell R, Alboji A, Haeussinger FB, Wurster I, Brockmann K, Suenkel U, Eschweiler GW, Maetzler W, Berg D, Fallgatter AJ. Age and vascular burden determinants of cortical **hemodynamics** underlying verbal fluency. PLoS ONE. 2015;10(9):e0138863.
2. Reed BR, Marchant NL, Jagust WJ, DeCarli CC, Mack W, Chui HC. Coronary risk correlates with cerebral amyloid deposition. Neurobiol Aging. 2012;33(9):1979–87.
3. Villeneuve S, Jagust WJ. Imaging vascular disease and amyloid in the aging brain: implications for treatment. J Prev Alzheimers Dis. 2015;2(1):64–70.
4. Mosconi L, Berti V, Glodzik L, Pupi A, De Santi S, de Leon MJ. Pre-clinical detection of Alzheimer's disease using FDG-PET, with or without amyloid imaging. J Alzheimers Dis. 2010;20(3):843–54.
5. Chetelat G, Desgranges B, de la Sayette V, et al. Mild cognitive impairment: can FDG-PET predict who is to rapidly convert to Alzheimer's disease? Neurology. 2003;60:1374–7.
6. Kogure D, Matsuda H, Ohnishi T. Longitudinal evaluation of earlier Alzheimer's disease using brain perfusion SPECT. J Nucl Med. 2000;41:1155–62.
7. Mosconi L. Glucose metabolism in normal aging and Alzheimer's disease: methodological and physiological considerations for PET studies. Clin Transl Imaging. 2013;1(4).
8. Joubert S, Gour N, Guedj E, Didic M, Guériot C, Koric L, Ranjeva JP, Felician O, Guye M, Ceccaldi M. Early-onset and late-onset Alzheimer's disease are associated with distinct patterns of memory impairment. Cortex. 2015;17(74):217–32.
9. Alsop DC, Detre JA, Grossman M. Assessment of cerebral blood flow in Alzheimer's disease by spin-labeled magnetic resonance imaging. Ann Neurol. 2000;47:93–100.
10. Johnson NA, Jahng GH, Weiner MW, et al. Pattern of cerebral hypoperfusion in Alzheimer disease and mild cognitive impairment measured with arterial spin-labeling MR imaging: initial experience. Radiology. 2005;234:851–9.
11. Xekardaki A, Rodriguez C, Montandon ML, Toma S, Tombeur E, Herrmann FR, Zekry D, Lovblad KO, Barkhof 615 F, Giannakopoulos P, Haller S. Arterial spin labeling may contribute to the prediction of cognitive deterioration in healthy elderly individuals. Radiology. 2014;274:490–9.
12. Chao LL, Buckley S, Kornak, J Schuff N. ASL perfusion MRI predicts cognitive decline and conversion from MCI to dementia. Alzheimer Dis Assoc Disord. 2010;24(1):19–27.
13. de la Torre JC, Mussivand T. Can disturbed brain microcirculation cause Alzheimer' disease? Neurol Res. 1993;15:146–53.
14. Gobel FL, Norstrom LA, Nelson RR, Jorgensen CR, Wang Y. The rate-pressure product as an index of myocardial oxygen consumption during exercise in patients with angina pectoris. Circulation. 1978;57(3):549–56.
15. Otto CM. Principles of echocardiographic image acquisition and doppler analysis. In: Otto CM, editor. Textbook of clinical echocardiography. 2nd ed. Philadelphia: WB Saunders; 2000. p. 1–29.
16. Hare JL, Jenkins C, Nakatani S, et al. Feasibility and clinical decision-making with 3D echocardiography in routine practice. Heart. 2008;94:440–5.
17. Ahmed SF, Syed F, Porembka D. Echocardiographic evaluation of hemodynamic parameters. Crit Care Med. 2007;35(suppl):S323–9.
18. Franklin KM, Aurigemma GP. Prognosis in diastolic heart failure. Prog Cardiovasc Dis. 2005;47:333–9.
19. de la Torre JC. Alzheimer disease prevalence can be lowered with non-invasive testing. J Alzheimer Dis. 2008;14:353–9.
20. Trojano L, Antonelli R, Acanfora D. Cognitive impairment: a key feature of congestive heart failure in the elderly. J Neurol. 2003;250:1456–63.

21. Zuccala G, Cattel C, Manes-Gravina E et al. Left ventricular dysfunction: a clue to cognitive impairment in older patients with heart failure. J Neurol Neurosurg Psychiatry. 1997;63:509–12.
22. Jefferson AL, Beiser AS, Himali JJ, Seshadri S, O'Donnell CJ, Manning WJ, Wolf PA, Au R, Benjamin EJ. Low cardiac index is associated with incident dementia and Alzheimer disease: the Framingham Heart Study. Circulation. 2015;131(15):1333–9.
23. Lewis TJ, Trempe CL. The End of Alzheimer's. 2014; pp 145–210.
24. Ascaso FJ, Cruz N, Modrego PJ, Lopez-Anton R, Santabárbara J, Pascual LF, Lobo A, Cristóbal JA. Retinal alterations in mild cognitive impairment and Alzheimer's disease: an optical coherence tomography study. Neurol. 2014;261(8):1522–30.
25. Lesage SR, Mosley TH, Wong TY, Szklo M, Knopman D, Catellier DJ, Cole SR, Klein R, Coresh J, Coker LH, Sharrett AR. **R**etinal microvascular abnormalities and cognitive decline: the ARIC 14-year follow-up study. Neurology. 2009;15, 73(11):862–8.
26. Qiu C, Fratiglioni L. A major role for cardiovascular burden in age-related cognitive decline. Nat Rev Cardiol. 2015;12(5):267–77.

Chapter 16
The Turning Point for Alzheimer's

How Prevention Works

There is general agreement that the ability of the brain to maintain normal cognitive function is dependent on optimal blood flow delivery to brain cells. The brain needs nutrients to carry out normal neurometabolism, neurotransmission, cognitive tasks, and conscious awareness. Lacking nutrients from insufficient cerebral blood flow results in a subnormal cerebral state vulnerable to neurodegeneration and dementia. This is a biological fact that is uncontestable.

Identifying many conditions that threaten the ability of blood vessels to provide the indispensable nutrients to neurons is one of the most important discoveries made in the field of Alzheimer's in recent times. These conditions are called **vascular risk factors to Alzheimer's dementia** (Figs. 12.1–12.3).

We have reviewed the role of vascular risk factors and their effect on the development of cognitive decline and Alzheimer's evolution in Chap. 12. The take home message about vascular risk factors from my point of view is that substantial evidence indicates they are precursors, together with advanced aging, of cognitive decline and Alzheimer's. The reason is that these Alzheimer risk factors add a critical burden to cerebral blood flow which normally declines progressively due to increased aging. Studies now indicate that vascular risk factors are already present in patients who develop mild cognitive impairment, a finding that places the presence of these factors long before substantial cognitive dysfunction and neurodegeneration develop [1]. We now suspect that mild cognitive impairment is a preclinical condition that bridges the stage between normal cognitive function and Alzheimer onset.

We have theorized in the past that in the presence of advancing age, vascular risk factors contribute to chronic brain hypoperfusion that progressively generates a neuronal energy crisis whose outcome is neurometabolic slowdown, senile plaque deposits, neurofibrillary tangle formation, synaptic failure, and brain cell death, mostly in cognitive-related regions (Fig. 7.1).

J.C. de la Torre, *Alzheimer's Turning Point*,
DOI 10.1007/978-3-319-34057-9_16

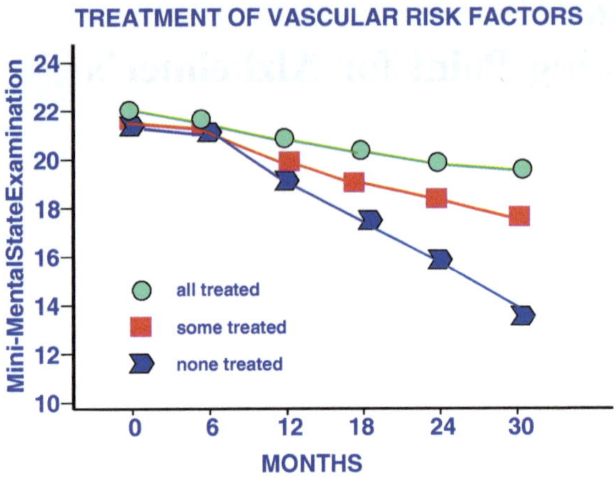

Fig. 16.1 Graphic shows improved mental status after 12-month treatment of patients with vascular risk factors in some (*red squares*) or all (*green circles*) diagnosed with Alzheimer's disease as compared to non-treated controls (*blue arrowheads*). The vascular risk factors treated included, diabetes type 2, high blood pressure, tobacco smoking, atherosclerosis, and elevated cholesterol levels. A Mini-Mental State Examination (MMSE) to measure neurocognitive ability was used to monitor cognitive decline in these three groups during a 30-month observation period. The higher the MMSE score, the less the cognitive decline; this is seen in the treated groups whose rate of cognitive decline was significantly less than non-treated patients (analysis of variance $p < 0.002$). Adapted from Deschaintre Y et al. Neurology 2009;73(9):674–80

Early management or treatment of such vascular risk factors such as diabetes type 2, high blood pressure, smoking, high serum cholesterol, and atherosclerosis has been reported to markedly slow down cognitive decline in Alzheimer patients [2] (Fig. 16.1). This is a good news because it indicates that severe cognitive dysfunction is preventable even when the red flags of memory complaints appear. Moreover, vascular risk factors offer a therapeutic target at an early stage to delay or prevent cognitive decline using proper treatment or lifestyle changes.

As promising as this sounds in battling Alzheimer's, there is one problem that needs to be resolved. If a middle age or older individual who appears cognitively healthy or shows only mild memory flaws, visits a cardiologist, internist, neurologist, or family practitioner for a nondescript complaint that has nothing to do with memory or heart disease, the patient will be given a prescription and sent home. **A missed opportunity is wasted**. No interest will be generated to seek more information regarding what other conditions may lurk in the brain or heart that could later affect cognitive function. There are exceptions to this rule, but they are just that, exceptions. It is a waste of time and effort to wring our hands or find excuses for the way medicine is practiced today in the USA because no amount of persuasion is going to change this well-established tradition.

Does that mean we are doomed by Nature's whim to become part of the Alzheimer epidemic at the age of 60 or 80? The short answer is yes. But like anything else that happens in the USA which eventually is considered appalling medical practice, there is a solution to markedly avoid the dementia fate. Some examples of egregious penchants that were changed with the idea of saving life and limb are smoking cessation, obesity, and case management for key chronic conditions. These three preventive measures have saved millions of lives, but there was a time when medical practice was lackadaisical in recommending life-saving changes to their patients.

To take a case in point, from the early 1940s, smoking was a way of life not only in the USA but in the entire civilized world. Smoking was considered glamorous by Hollywood standards, adventurous by the Marlboro man, relaxing by the connoisseurs, and as pleasurable and safer than as sex by cigarette company advertisers. Little was known about the 4000 chemicals packed in a cigarette, 69 which were carcinogenic and another 250 toxins known to cause other diseases. Doctors were paid by cigarette companies to advertise their products as safe which they happily did despite strong evidence at the time that smoking could kill.

In the late 1920s, German scientists were the first to identify the link between smoking and lung cancer, but by the start of World War II, this link was ignored and forgotten in the interests of military pursuits. In 1950, a cause–effect relationship between smoking and lung cancer was confirmed in a landmark study by Wynder and Graham [3]. However, it took another 14 years for the US Surgeon General's Report in 1964 to start the long process of educating the public and conservative physicians that smoking tobacco was bad for you. You did not even have to smoke; second-hand inhaling tobacco from a smoker causes 34,000 deaths in the USA from heart disease and lung cancer. Despite these statistics, the US Congress has refused to pass any federal law banning smoking from public places and leaves the responsibility to the states and the consumer. Such a ban would deprive $32 billion in tax dollars to state governments, a considerable income that seems to trump public health. The tobacco industry lobbies Congress aggressively with serious money and dubious rhetoric that smoking is an individual freedom. This tactic ensures maintaining a large customer base, stable markets, and higher profits.

Statistics show that cigarette smoking causes 90 % of lung cancer deaths and that not smoking can significantly reduce not only the risk of other cancers, but also heart attacks, stroke, and chronic lung disease. While chemotherapy, surgery, and irradiation are last ditch efforts for treating lung cancer, there is no argument contradicting that without prevention, this lethal disorder would be out of control. There is now convincing evidence that chronic smoking increases the risk of Alzheimer's [4].

A second mechanistic example of prevention with wide medical implications is vehicular accidents. Prior to 1965, there were an estimated 50,000 people who died annually in automobile accidents and an equal number who sustained severe neurological deficits. While the search for drugs and new surgical techniques to repair the carnage caused by vehicular accidents was in full swing, it was not until the US Congress passed the Highway Safety Act in 1966 that required cars be equipped with safety belts, energy absorbing steering wheels, shatter-resistant windshields, head rests, and later air bags, that dramatically changed the death and injury statistics, which by some estimates may have saved millions of lives.

The Rising Alzheimer's Incidence

The somber concern generated by the rising incidence of Alzheimer's disease is that it has reached critical mass and is now at the epidemic threshold of its limits. If allowed to continue on its present growing trend, it is anticipated that the number of new Alzheimer cases in the USA alone will balloon from 5.3 million people currently affected to nearly 14 million people by 2050 and will nearly quadruple from 35 million people affected worldwide to 140 million (Fig. 8.1). This calamitous escalation of Alzheimer's can be partly attributed to the rise of the baby boomer population aged 65 and over which will triple in the USA by 2050.

However, and possibly more important than the escalating elderly population is the fact that for the last 20 years there has been an emphasis in the search for drugs that can 'cure' Alzheimer's. Unfortunately, the brains of Alzheimer patients at a moderate to late stage are already devastated from neurodegeneration beyond repair. This failed pie-in-the-sky strategy by mercenary pharmaceuticals has ignored the principles of disease prevention which state that prevention needs to be applied before the pathogenesis and irreversibility of a disorder have evolved. Disease prevention is generally possible when a mechanistic understanding of the pathogenic process is understood and treatment is applied to quell or reverse the pathogenic process. If poor blood flow is the mechanistic process behind the evolution of cognitive decline, and substantial evidence indicates this is the case, efforts should be made to detect and prevent the factors that contribute to this pathologic process. A key point in understanding how vascular risk factors may impact cognitive function is their association with a dynamic reduction in cerebral blood flow [5]. The reduced blood flow to the brain mainly originates from three sources, the heart, the periphery and the brain. The reduction in blood flow to the brain from vascular risk factors excludes non-modifiable risk factors to Alzheimer's involving age, gender, genetics, education, ApoE4 allele, and a prior brain injury, any of which can add an extra burden on blood flow delivery to the brain.

Clearly, the field of medicine needs to formulate a radical plan to meet the Alzheimer challenge of substantially slowing down the number of new cases that are bound to overwhelm the medical facilities now available. In my judgment, there is no need to re-educate physicians as to their responsibility according to the modern version of the Hippocratic Oath which states "I will prevent disease whenever I can." Rather, what is needed is a multidisciplinary center dedicated to Alzheimer's disease prevention, much like a Children's Hospital or a Cancer Center which focus on a particular population and enjoys an experienced staff of specialists and support personnel.

There is hope that prevention can significantly lower the incidence of Alzheimer's in years to come. But, how to best accomplish this? A substantial number of studies now indicate that identifying and detecting the major vascular risk factors associated with Alzheimer's before its onset can provide early preclinical interventions to better prevent progressive cognitive failure [6] (see Figs. 12.1–12.3, 14.3 and 16.1). Alzheimer's prevention at its earliest stage will be difficult without the creation of dedicated multidisciplinary centers as described here.

Heart-Brain Clinics Can Lower the Number of New Alzheimer Cases

To optimize detection, prevention, and preclinical treatment of Alzheimer's risk factors before irreparable cognitive loss occurs, **requires the establishment of outpatient Heart-Brain Clinics in major American cities**. This venture will need a significant funding investment by the federal government initially, but it will yield a high return in 6 major areas:

1. it will substantially reduce the number of new Alzheimer and dementia cases at an early preclinical stage through preventive measures;
2. it will substantially reduce societal and Medicare costs estimated to top one trillion dollars by 2050;
3. by establishing and maintaining a large data base of its activities, it will advance our knowledge of dementia with the aim of **reversing** cognitive loss;
4. it will provide a database that can be used to assess efficacy and safety of treatments and management of patients;
5. It will reduce errors of diagnosis using state-of-the-art equipment, special testing and experts to run them;
6. It will personalize treatment or management of patients that improves mental outcome.

The main goal of the Heart-Brain Clinics would be to detect and prevent vascular risk factors to Alzheimer's, ideally before cognitive impairment signs appear. The second goal of the Heart-Brain Clinics would be to treat or manage active cognitive impairment in order to contain progressive mental deterioration.

When vascular risk factors are detected, a plan for appropriate intervention or management can be applied (Fig. 16.2). When this plan is properly followed, it will extend quality of life, physical health, and cognitive function.

Because of the close physiopathological relationship that many dementias have with each other, **the Heart-Brain Clinics do not need to limit themselves to Alzheimer's prevention but can extend the same parameters used to detect Alzheimer's to such disorders as vascular dementia, frontotemporal dementia, and dementia with Lewy bodies**. Differential diagnosis to detect these non-Alzheimer's dementias is useful since their treatments and management are comparable to those that will be found useful for Alzheimer's.

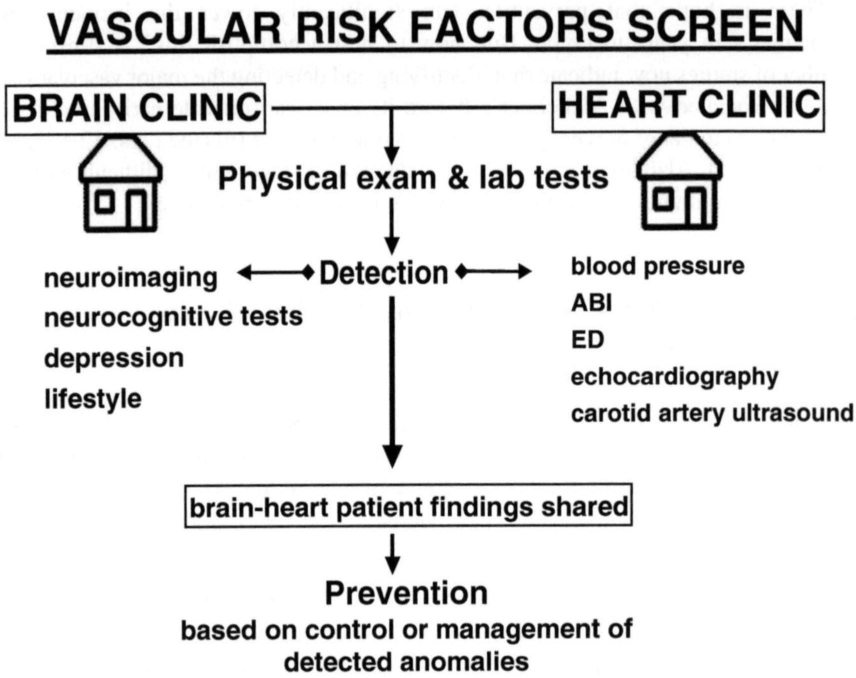

Fig. 16.2 Simplified plan to rule out vascular risk factors associated with Alzheimer's disease in an outpatient **Heart-Brain Clinic** using noninvasive, cost-effective tools. Screening of individuals is based on medical referrals to the Heart-Brain Clinic. Following a physical exam and routine laboratory blood tests, patient history or current complaint determines the sequence of brain and heart testing. The multidiagnostic findings from heart and brain testing are reviewed by the attending physician, and a medical plan of action or follow-up observation period is determined. Periodic follow-up visits by the patient determine any further changes in the plan of action or in continued observation. The Heart-Brain Clinic's main goal is to detect vascular risk factors to Alzheimer's ideally before cognitive impairment appears or when daily living activities can be carried out by the patient. This approach can extend quality of life, physical health, and preserve cognitive function as long as possible. Key: *ABI* ankle–brachial index, *ED* erectile dysfunction questionnaire

In order to streamline a cost-effective, pragmatic plan to screen vascular risk factors that can promote incident Alzheimer's, Heart-Brain Clinics need to be integrated under one specialized center staffed by a dedicated **multidisciplinary team made-up of neurologists, psychologists, cardiologists, neuroradiologists, and technical support personnel**. Once operational, the goal of these centers is to provide a more tailored and earlier clinical assessment of cognitive function that will result in a much improved patient outcome.

Improved patient outcome implies risk stratification, the process of separating high and low risk of Alzheimer's disease, as well as targeted medical decision making and tailored intervention following diagnostic findings from each Heart-Brain Clinic for each patient examined. This plan can provide an appropriate and personalized outpatient management or modifying intervention, assuming one is available (Fig. 16.2). An additional advantage of a multidisciplinary approach is that modifying interventions can be applied to the patient earlier in the development of cognitive decline where time may be a deciding factor in the eventual reversibility of any ongoing cognitive pathology.

Moreover, pooled patient diagnostics from both brain and heart clinical findings are less likely to misdiagnose Alzheimer's when dyscognitive symptoms are due to other causes, such as normal pressure hydrocephalus, medications, depression, thyroid disease, or vitamin B12 deficiency. Finally, patients seen at Heart-Brain Clinics are less likely to be lost to follow-up than with monotherapeutic practice.

Once established, patients learn of these specialized Heart-Brain Clinics from news sources, advertising, or from primary care facilities, which can refer patients to these centers. Heart-Brain Clinics are presently not part of medical practice in the USA although it is within the means of most level 3 hospitals. Establishing a Heart-Brain Clinic in major US cities can be done rather quickly once the funding is secured by adding a center to an already operating level 3 hospital which is equipped to handle all services. Heart-Brain Clinics would not admit patients but serve simply as an outpatient diagnostic and treatment center for the prevention or delay of cognitive disorders that can lead to dementia.

A similar plan from the one suggested here is presently serving patients at the VAS-COG Clinic in Florence, Italy, under the direction of **Dr. Leonardo Pantoni** [7]. A heart-brain clinical program may soon be introduced by a team headed by **Professor Mat Daemen** in the Netherlands [8]. The focus of these centers is on dementia prevention. It is not the aim of heart-brain clinics to systematically screen a chosen population group such as the elderly, but rather to screen people at risk of cognitive deterioration following referrals or patient walk-ins. Such risk is determined by healthcare personnel that can identify people with subtle, mild, or moderate cognitive complaints. The essential goal of Heart-Brain Clinics is earliest detection of Alzheimer vascular risk factors that can be managed, treated, or prevented to alter the course of potential Alzheimer onset.

It is assumed that if successful, this plan can be applied to other countries around the world. Although the Heart-Brain Clinics are best staffed by specialists with training in neurology, cardiology, psychology, and neuroradiology, these specialists need to be trained in all aspects of cognitive diagnostics and treatment options as well as being cognizant of the pathophysiology involved in dementia. That implies that physicians with training in other branches of medicine can also be trained to recognize the principal signs of potential or actual cognitive deterioration in an older patient and order the necessary tests that would help with the diagnosis.

The central idea of gathering trained specialists in neurology, cardiology, psychology, and neuroradiology under one roof is that diagnostic findings for each patient will be evaluated to examine three main systems in the body that can reflect impending or actual cognitive failure. These three systems are the brain, the heart, and the mind. These systems and the potential detection of vascular risk factors which they are capable of generating is a key element that opens the door for an effective means to combat this dementia.

An example of this strategy would work as follows. Let us say a 52-year-old patient walks-in at a Heart-Brain Clinic complaining of light-headedness and mild memory difficulties. Following the plan in Fig. 16.2, a physical exam is performed and routine laboratory tests ordered. Assuming no red flags are found, the patient undergoes neuroimaging (e.g., MRI-ASL) and neurocognitive tests. Neurocognitive tests show very subtle cognitive deficits and neuroimaging reveals a relative decrease in cerebral blood flow in the posterior cerebral cortex. The patient is now sent to the cardiology section of the clinic for cardiac evaluation. An echocardiogram discloses paroxysmal atrial fibrillation (an Alzheimer risk factor). This common cardiac arrhythmia comes and goes and can remain asymptomatic. The patient is given rate control therapy, and a follow-up visit 3 months later shows relief of light-headedness and neurocognitive improvement over baseline tests.

Heart-Brain Clinics will eventually replace centers that practice monotherapeutic diagnostics of Alzheimer's disease, including memory units in hospitals or universities as well as geriatric clinics whose focus is mostly on elderly individuals. The goal of the Heart-Brain Clinics will be to apply and interpret noninvasive, cost-effective, multidiagnostic testing of heart and brain functions in an outpatient setting for asymptomatic and symptomatic patients at risk of dementia. In medicine, usually only one specialist is required to diagnose and treat a patient with cardiac disease, high serum cholesterol, memory problems, diabetes, or stroke. Alzheimer's is different because many conditions, such as those just mentioned, can promote the start of cognitive deterioration in middle-aged or advanced-aged individuals. That means that if a specialist in cardiac disease or diabetes for example is confronted with a patient showing signs of cognitive loss, a referral to a neurologist or psychologist is generally indicated to provide an appropriate treatment. The same logic applies if a psychologist or a neurologist examines a patient that shows signs of cognitive deterioration, is it due to stroke, diabetes, heart disease or some other condition?

Heart-Brain Clinics require an initial funding of $10 billion dollars. This is a cost-effective estimate considering Alzheimer's cost to society in the USA is anticipated to explode yearly from a present $228 billion to over a trillion dollars by 2050 (source, Alzheimer's Association). This sum can initially create 8–10 Heart-Brain Clinics fully staffed in selected American cities. It is recommended that trial Heart-Brain Clinics be established in select US cities so their merit in markedly slowing or preventing incident Alzheimer's can be compared to the present monotherapeutic approach. In addition, Heart-Brain Clinics can provide a more complete healthcare assessment of its patients by coincidental discovery of other disorders unrelated to dementia such as neurological abnormalities and non-vascular cardiac disease.

Following their anticipated effectiveness in lowering Alzheimer's incidence, additional centers can be established in other cities. Funding for Heart-Brain Clinics can be raised from government, capital investments, public contributions, and industry. The outcome of such an enterprise is healthier living during advanced aging and less chance that dementia will bankrupt Medicare and Medicaid.

Challenges in Establishing Heart-Brain Clinics

Since no strategic plan to prevent the rising tide of Alzheimer's is perfect, some challenges in establishing Heart-Brain Clinics may emerge:

1. Heart-brain clinics require secure funding to staff, properly equip, and run efficiently;
2. Recruited specialists and technical personnel must be able to communicate with each other concerning patient findings;
3. Understanding and dealing with asymptomatic or preclinical dyscognitive patients requires acquiring expertise in the field of neurocognitive behavior;
4. Excellent organizational and managerial leadership is crucial for the operational success of Heart-Brain Clinics;
5. Inadequate patient's health insurance or affordability of services could limit access to Heart-Brain Clinics.

Although a considerable number of memory clinics and cardiac clinics have proliferated in the last two decades in the USA and Europe to accommodate the rising elderly population, these clinics are not organized to exchange patient findings. Specialty clinics simply ignore each other. The trick, therefore, is for brain and heart specialists to collaborate with each other under the same roof and determine a more coherent plan of action for patients likely to initiate or worsen dyscognitive behavior in an outpatient setting. This action requires organization and the provision of guidelines by the hospital or health center administrators to educate its physician heart-brain staff in the rewarding task of interpreting, integrating, and sharing patient findings that can vastly improve patient outcome. This is not how such patients are handled under the present system in the USA, or elsewhere in the world. This lack of action is a wasted opportunity for safeguarding patient wellness.

It is important to recall that in science very few things that occur in nature can be considered a certainty. Following this maxim, medical knowledge is always ready to be modified and this rule applies to the understanding of human cognitive behavior. The proposed Heart-Brain Clinic approach is not perfect, but it offers for the first time, a clear path to defeat this dreaded disorder before it rears its ugly fangs. This is as good as it gets for many generations to come.

The three major advantages of establishing Heart-Brain Clinics to assess patients at risk of Alzheimer's are:

1. Better risk stratification and medical decision making than a monotherapeutic approach;
2. Earlier detection of predisposing vascular risk factors to Alzheimer's by examining two essential organ systems, heart and brain;
3. Tailoring of intervention or management by focusing on specific or potential abnormalities found after patient vascular risk factors screening.

In the past, I have summarized my medical philosophy in a breviloquent precept:

The goal of medicine is to provide patients with hope, and when there is no hope, to offer understanding.

I am convinced that we stand at the threshold of eliminating Alzheimer's and other dementias as neuropathological curses of aging. I see this as preventive measures become more specific and practical in the future. Treatment of vascular risk factors can currently lower cognitive decline in people diagnosed with Alzheimer's (Fig. 16.1). For now, prevention by detection and treatment of modifiable vascular risk factors must be considered as the most judicious plan of action and one that will pave the way to a life unencumbered by the loss of our consciousness.

A healthy and happy life can be had to age 100 and beyond. A number of studies have shown that people who reach the centenarian age do not usually die of 'old age' or Alzheimer's disease but rather of common diseases such as cancer or heart attack [9]. Studies of centenarians indicate that if cognitive deterioration can be dodged or delayed by some means, a long life can be had without the fear of cognitive meltdown. The people who achieve this are known as "delayers" or "escapees." Delayers are those that have somehow postponed mental deterioration by receiving interventions, following a healthy lifestyle, or inheriting favorable genetics. Escapees are a rare bunch of people that have few or no life-threatening diseases by the age of 90 or beyond.

Medical technology is presently in a position to offer elder individuals a delay from cognitive turmoil by applying detection and preventive measures as discussed above. To successfully escape from dementia, the delaying measures essentially need to take a giant leap forward technologically, a development involving much improved detection of Alzheimer precursors and safely boosting brain blood flow that essentially neutralizes cerebral hypoperfusion and its consequences [10].

We need to move with a sense of urgency with respect to establishing Heart-Brain Clinics because this is a real opportunity to provide a significant difference in extending quality of life rather than the specter of Alzheimer's disease.

References

1. Tervo S, Kivipelto M, Hanninen T. Incidence and risk factors for mild cognitive impairment: a population-based three-year follow-up study of cognitively healthy elderly subjects. Dement Geriatr Cogn Disord. 2004;17(3):196–203.
2. Deschaintre Y, Richard F, Leys D, Paquier F. Treatment of vascular risk factors is associated with slower decline in Alzheimer's disease. Neurology. 2009;73:674–80.
3. Wynder EL, Graham EA. Tobacco smoking as a possible etiologic factor in bronchiogenic carcinoma: a study of six hundred and eighty-four proved cases. JAMA. 1950;143:329–36.
4. Anstey KJ, von Sanden C, Salim A, O'Kearney R. Smoking as a risk factor for dementia and cognitive decline: a meta-analysis of prospective studies. AmJEpidemiol. 2007;166(4):367–78.

5. Zhang S, Smailagic N, Hyde C, Noel-Storr AH, Takwoingi Y, McShane R, Feng (11)C-PIB-PET for the early diagnosis of Alzheimer's disease dementia and other dementias in people with mild cognitive impairment (MCI). J Cochrane Database Syst Rev. 2014;7: CD010386.
6. de la Torre JC. Alzheimer's disease: detection, prevention, and preclinical treatment. J Alzheimers Dis. 2014;42(Suppl 4):S327–8.
7. Ciolli L, Poggesi A, Salvadori E, Valenti R, Nannucci S, Pasi M, Pescini F, Inzitari D, Pantoni L. The VAS-COG clinic: an out-patient service for patients with cognitive and behavioral consequences of cerebrovascular diseases. Neurol Sci. 2012;33:1277–83.
8. van Buchem MA, Biessels GJ, Brunner la Rocca HP, de Craen TJM, van der Flier WM, Ikram MA, Kappelle J, Koudstaal PJ, Mooijaart SP, Niessen W, van Oostenbrugge R, de Roos A, van Rossum B, Daemen MJAP. The heart-brain connection: a multidisciplinary approach targeting a missing link in the pathophysiology of vascular cognitive impairment. J Alzheimers Dis. 2014;42(Suppl 4):S443–51.
9. Motta M, Bennati E, Vacante M, Stanta G, Cardillo E, Malaguarnera M, Giarelli L. Autopsy reports in extreme longevity. Arch Gerontol Geriatr. 2010;50(1):48–50.
10. Chen Y, Wolk DA, Reddin JS, Korczykowski M, Martinez PM, Musiek ES, Newberg AB, Julin P.

Chapter 17
How Poor Brain Blood Flow Promotes Alzheimer's Disease

Brain Blood Flow and Metabolism

The central thesis of this book is that Alzheimer's is induced from poor blood flow to the brain. Because neurodegeneration and cognitive loss take many years to manifest from the time when poor blood flow to the brain is first detected, Alzheimer's is ostensibly preventable.

To take a common example of insufficient blood flow to an organ, coronary artery narrowing that produces chest pain, shortness of breath, and lightheadedness can be detected before any of these symptoms appear using **coronary computed tomography angiography** which can quantify non-invasively coronary atherosclerosis. Thus, the outcome of coronary artery disease can be prevented at an early stage before prognosis worsens. So it is with Alzheimer's disease. A number of screening tests reviewed in Chap. 15 can clearly prevent or delay cognitive deterioration if a major plan is designed for this purpose (for the plan, see Fig. 14.3 and for discussion, see Chap. 16, Heart-Brain Clinics).

This vasculogenic pathology that induces cognitive loss and severe brain damage is essentially based on how blood flow behaves as we age and how it can misbehave into devastating large portions of the brain. The physiological process leading to a state of cognitive failure is difficult to understand without a basic grasp of how critical the role of brain blood flow and hemodynamic disturbances has on brain function.

The concept that brain blood flow is tightly associated with how we think, communicate, and maintain consciousness may be credited to hundreds of researchers in the nineteenth and twentieth century. In particular, two British investigators, **Drs. Charles Sherrington** and **Charles Roy**, wrote a seminal paper in 1890 on the regulation of blood flow to the brain. They described the now classical concept that "*…the brain possesses an intrinsic mechanism by which its*

J.C. de la Torre, *Alzheimer's Turning Point*,
DOI 10.1007/978-3-319-34057-9_17

vascular supply can be varied locally in correspondence with local variations of functional activity."

The idea that energy supply (blood flow) is controlled by energy demand (brain metabolism) was revolutionary at the time and is no less stunning at the present because it provides a glimpse of how our brain works and how the environment or disease can affect the supply of blood flow available to feed the neurons that make us who we are. Although the Sherrington-Roy idea oversimplifies the complex workings of cerebral blood flow and cognitive behavior, at the very least it provided a useful account that could help explain how abnormal physiological states affect the aging or diseased brain. The Sherrington-Roy description of blood flow and metabolic activity is now known as **neurovascular coupling**. Neurovascular coupling can be shown using high-resolution neuroimaging techniques such as functional magnetic resonance imaging (fMRI) or fluorodeoxyglucose positron emission tomography (FDG-PET) where brain blood flow response to local brain activity can be visualized and measured. For example, recognizing a picture of someone we know will activate neurons in a specific region of the brain which will increase local blood flow in that region. Brain neurons are highly sensitive to stimuli that activate them, and a corresponding change in blood flow is at their command. If you tap your index finger against your thumb repeatedly, blood flow will increase 60 % to the neurons of the motor cortex that control that activity.

Arterial spin labeling fMRI is a neuroimaging technique that can measure neuronal activity by detecting changes in the oxygenation of the blood, a method called BOLD (blood oxygenation level dependent). Since blood oxygenation varies according to neuronal activity, BOLD is used to detect brain activity in a direct manner. This has been used to measure brain blood flow changes that occur in people at risk of Alzheimer's or during onset of mild cognitive impairment. As the technology of these neuroimaging techniques improves with time, it is not far-fetched to expect highly accurate and reliable brain activation maps of patients at risk of incipient cognitive dysfunction years before mental changes are manifested. The patient's history, lifestyle, and vascular risk factors can then be surveyed in order to manage or treat potentially harmful elements that can compromise the patient's health. Findings using these neuroimaging techniques have already produced valuable information of how memories, language, anxiety, and learning are formed in normal and abnormal settings.

When Blood Flow to the Brain Is Insufficient

Any event that disrupts blood flow delivery to the brain results in loss of cognitive function and consciousness within seconds, and if that disruption is maintained for 4–5 min, irreparable brain damage or death will occur. Blood flow provides vital oxygen and glucose to create energy for brain cells without which these cells do not survive. The brain does not store or produce oxygen and glucose so it is totally dependent on these high-energy nutrients to be delivered by the circulation in a timely and sufficient manner.

Restricting either oxygen or glucose can result in loss of consciousness or spell disaster for brain cells. This outcome can occur when oxygen or glucose is unavailable as it can happen when choking on food or during a hypoglycemic attack from insulin overdose. Quick restoration of deficient oxygen or glucose is unlikely to harm the brain. Normal blood flow to the brain in adult humans is about 60 ml (expressed as milliliters per 100 g of tissue per minute). Regionally, flow is about 70 ml in gray matter and 20 ml in white matter. These values normally decrease with age by an estimated rate of about 0.5 % per year [1]. Between the ages of 20 and 65, normal CBF generally declines about 15–20 % [2]. This drop in blood flow does not pose a problem during aging unless blood flow drops to a critical level from the presence of vascular risk factors (Figs. 12.1–12.3) or poor health. It is generally accepted that when cerebral blood flow reaches 30 ml, neurologic symptoms appear, and when it drops to 15–20 ml, electrical failure or irreversible neuronal damage can occur even within minutes.

The capacity to maintain an adequate and stable cerebral blood flow in the face of widely changing mean blood pressures ranging from a high of 150 mm Hg to a low of 60 mm Hg is dependent on **cerebral autoregulation**. This is a process that kicks into dilate or constrict small arteries and arterioles in the brain when complex physiological control systems sense hyper- or hypoperfusion.

When mean blood pressure drops below the limit of cerebral autoregulation, ischemia and mental changes will result. Experiments have shown that autoregulation is lost after a severe head injury or space occupying brain mass such as a tumor or hematoma. It is less clear how autoregulation works when chronic brain hypoperfusion is present for years during advanced aging. It is suspected there may be a point in time when cerebral autoregulation ceases to work when brain blood flow is unable to fully meet metabolic demand.

What happens when cerebral blood flow is unable to meet neuronal demand? This can occur when the heart is not pumping sufficient blood to the brain or when vessels supplying the brain encounter resistance from vessel stiffness, narrowing, or blockade. In the brain, blood flow is dependent on the resistance posed by the arterioles and capillaries (Fig. 17.1).

Vessel resistance is affected by three factors:

(1) the diameter of the vessel lumen is the most important factor for regulating blood flow within an organ, as well as for regulating arterial blood pressure. Changes in vessel diameter in small arteries and arterioles enable the brain to adjust their own blood flow to meet the metabolic requirements of the tissue.
(2) the viscosity of the blood (the higher the viscosity, the greater the resistance) also affects resistance. However, viscosity normally stays within a small physiological range, except in rare hematocrit disorders or when red blood cells are unable to deform into the shape of a bullet in order to flow through tight capillary lumens.

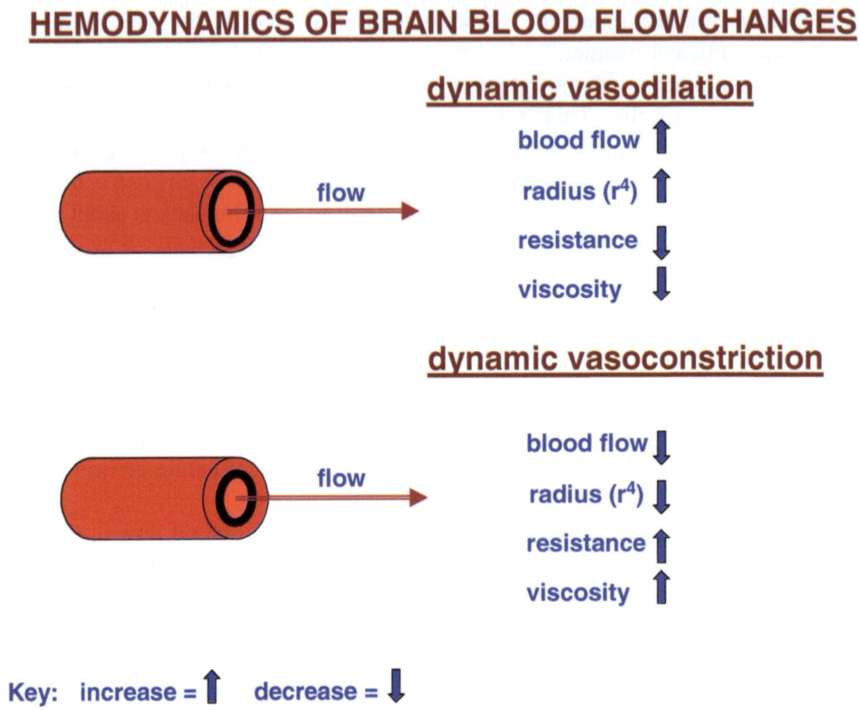

Fig. 17.1 Hemodynamic changes in a precapillary arteriole after vasodilation and vasoconstriction. Vasodilation occurs when smooth muscle cells (SMC) that surround the arteriolar wall relax; vasoconstriction involves contraction of the SMC. Small arterioles control the flow of blood into the capillaries, site of the blood–brain barrier, where oxygen, vital nutrients, and the removal of carbon dioxide (CO_2) take place (not shown). The dynamic balance between vasodilation and vasoconstriction in the brain can be altered depending on the physiological and pathophysiological conditions present. For example, hypertension decreases after vasodilation while exposure to cold induces the opposite reaction

(3) the length of the vessel can influence blood flow; the longer the vessel, the greater the resistance and lesser the blood flow. However, the vessel length does not vary significantly in normal physiology (Fig. 17.1).

The diameter or radius of the arteriolar vessel can change in response to signals from activated neurons to either relax or tighten the vascular smooth muscle layer that surrounds the outer wall of the vessel. This change can allow more blood to flow to neurons through *vasodilation* or less blood flow through *vasoconstriction*. A rather simple equation involving vessel diameter, length, and viscosity of fluids within a tube was first described by the nineteenth-century French physician **Jean Lèonard Marie Poiseuille**.

Poiseuille's Formula

What is extraordinary about Poiseuille's equation is that it touches on two crucial and fundamental issues relevant to cognitive deterioration. First is the dramatic effect that very small changes in the cerebral vessel's diameter can have on resistance and on the rate of blood flow to the brain as well as within the brain. **Poiseuille observed that blood flow is proportional to the 4th power of the vessel diameter**. This has monumental significance for brain function because **if the local vessels feeding a neuronal population are reduced by only 16 %, blood flow will be cut by half**. By the same token, if vessel diameters can be increased from their normal state by 19 %, blood flow will double. This effect on blood flow assumes that the applied pressure on these vessels stays the same.

The driving pressure for blood flow in the brain is the **pressure gradient where blood in an artery or arteriole flows between the points of highest to lowest pressure**. Cerebral blood flow is dependent on the pressure gradient established between arteries and veins. For example, assuming the systemic circulation is a long tube looping from the heart to the brain and back, if the mean pressure in the aorta (or inlet) is 90 mm Hg and ends at the junction of the vena cava with the right atrium (or outlet) where the mean pressure is 2 mm Hg, the mean pressure gradient

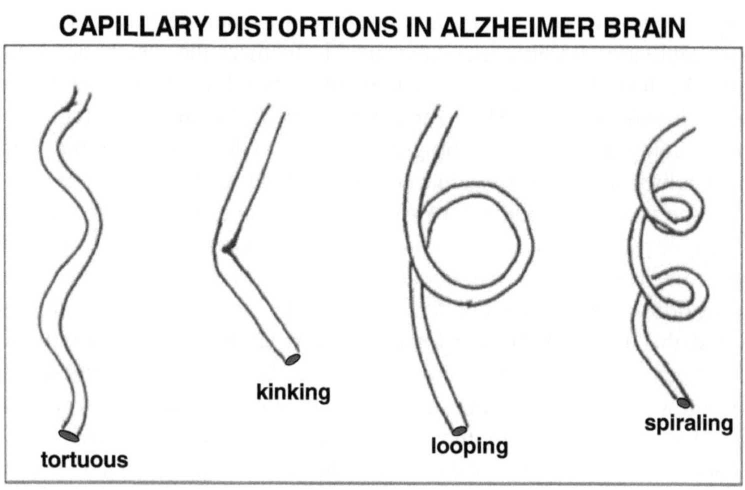

Fig. 17.2 Four types of capillary or arteriolar distortions that can be found in Alzheimer's brain. Such distortions can induce abluminal and intraluminal changes which markedly distort their normally cylindrical shape. Applying basic laws of fluid dynamics to blood flowing through these deformed microvessels, the following conditions can occur: **a** *flow velocity* is decreased; **b** *disturbed flow* or *microturbulence* develops; **c** capillary *resistance* to blood flow is increased; **d** blood *viscosity* is increased. Any or several of these conditions will result in less blood flow delivery to neurons whose survival may be compromised. Adapted from H.C. Han, J. Vasc Res. 2012;49: 185–197

in the tube is 88 mm Hg. If there is no pressure gradient in a vessel from inlet to outlet, there will be no flow in that vessel.

In our original thinking about the Poiseuille formula, it seemed evident that inadequate cerebral blood flow resulted from tortuous, kinking, or looping capillaries, as Fischer had found in the brains of Alzheimer's patients [3]. These small vessel distortions will upset the supply and demand homeostasis established by blood flow and brain cell metabolism. Hemorheological laws state that fluid passing through a tube, whose lumen is either kinked, irregular, or looping, will likely produce a flow whose pattern will be uneven and disturbed (Fig. 17.2). The disturbed flow is bound to exert more friction on the inner vessel wall which in the case of arterioles is bound to damage the endothelial cells, thus diminishing the endothelial vasoactive molecules that regulate blood flow across those vessels. This clinical picture can worsen when vascular risk factors are present because these factors lessen the blood flow available to the brain, a deficit often seen during advance aging.

Making Sense of the Vascular Hypothesis of Alzheimer's Disease

The vascular hypothesis of Alzheimer's that we proposed in 1993 was based on the concept that chronic brain blood flow insufficiency resulting from distorted arterioles and capillaries became inevitably unable to meet the metabolic demand of neurons. We had shown this to occur in old rats whose brain blood flow was reduced experimentally by 35 % for 3–6 months. This amount of reduced blood flow in the aging rats resulted in progressive memory loss, neurochemical changes, hypometabolism, and dead or damaged hippocampal neurons [4].

But, if the vascular hypothesis of Alzheimer's were to make sense, other factors besides capillary or arteriolar distortions needed to contribute to cognitive decline and dementia. The reason is that not all Alzheimer's brains examined contain these vessel distortions. This fact put a wrinkle in the vascular concept until it was discovered that many Alzheimer's patients were dying with comorbid disorders such as hypertension, atherosclerosis, cardiovascular disease, and type 2 diabetes, to name a few. It became evident that these conditions (or risk factors) could reduce cerebral blood flow, especially in older individuals who were already burdened by age-related poor blood flow to the brain. The rubric that vascular risk factors in older individuals led to an increased probability of acquiring Alzheimer's, now made more sense, and this was duly noted in my 1997 paper [5].

Dr. Cristina Polidori, a neurology professor at the University Hospital of Cologne, kindly pointed out to me that in her neurological practice, roughly 10 % of her mild cognitive impaired patients, a condition that can lead to Alzheimer's, show no signs of vascular risk factors and asks how this might occur. My response is that by the age of 50, it is unusual for an individual to be free of vascular risk factors or microangiopathic changes as described in Fig. 17.1. Moreover, the present diagnostic technology for measuring cerebral blood flow and hemodynamic changes is presently quite good but not perfect. For example, there can be instances when small hypoperfused brain regions may be missed in the neuroimaging examination, a detail that is surely to be corrected in the near future as the imaging technology rapidly improves.

The National Alzheimer's Coordinating Center reported [6] that post-mortem examination of 4629 individuals who died with an Alzheimer's diagnosis revealed vascular brain pathology in 79.9 % of all cases. That leaves about 20 % of the remaining cases unaccounted for other vascular risk factors including cardiovascular and peripheral artery disease that can influence cerebrovascular pathology by chronic brain hypoperfusion, a condition not readily detected by gross pathologic examination.

However, if indeed Alzheimer's can arise without any vasculopathic component driving the dementia, then those people free of vasculogenic complications should be carefully examined to see what other events could have spurred cognitive loss.

The role of blood flow in the brain is fundamental to understanding how cognitive decline is initiated in the elderly individual. This understanding is not simply academic but is directly relevant to potential therapy that can prevent or delay cognitive meltdown. The field of Alzheimer's has been distracted for several decades with what I call "intellectual back scratching" of theories financially supported by big pharma that do not medically help patients avoid or ease the suffering of Alzheimer's. These theories are no longer even considered contentious by some but outright obstructions to research progress. For that reason, we need to consider moving on with a sense of urgency toward preventing this mind-robbing disorder. Brain hypoperfusion in patients at risk of Alzheimer's has become a realistic target to achieve even though raising cerebral blood flow by direct intervention is at its infancy in terms of what can be done right now. Nevertheless, in the interest of shaking the tree to see whether an interesting notion falls in our hands, a review of these potential interventions is presented in the chapter that follows.

References

1. Leenders KL, Perani D, Lammertsma AA, Heather JD, Buckingham P, Healy MJR, Gibbs JM, Wise RJS, Hatazawa J, Herold S, Beaney RP, Brooks DJ, Spinks T, Rhodes C, Frackowiak RS, Jones T. Cerebral blood flow, blood volume and oxygen utilization. Brain. 1990;113:24–47.
2. Chen JJ, Rosas HD, Salat DH. Age-associated reductions in cerebral **blood** flow are independent from regional atrophy. Neuroimage. 2011;55(2):468–78.
3. Fischer VW, Siddigi A, Yusufaly Y. Altered angioarchitecture in selected areas of brains with Alzheimer's disease. Acta Neuropathol. 1990;79:672–9.
4. de la Torre JC, Fortin T, Park G, Butler K, Kozlowski P, de Socarraz H, Saunders J. Chronic cerebrovascular insufficiency induces dementia-like deficits in aged rats. Brain Res. 1992;582:186–95.
5. de la Torre JC. Hemodynamic consequences of deformed microvessels in the brain in Alzheimer's disease. Ann NY Acad Sci. 1997;26(826):75–91.

6. Toledo JB, Arnold SE, Raible K, Brettschneider J, Xie SX, Grossman M, Monsell SE, Kukull WA, Trojanowski JQ. Contribution of cerebrovascular disease in autopsy confirmed neurodegenerative disease cases in the National Alzheimer's Coordinating Centre. Brain. 2013;136(Pt 9):2697–706.

Chapter 18
Interventions that May Increase Cerebral Blood Flow

Photobiomodulation to Enhance Neuron Energy Fuel

The logic of treating vascular risk factors to prevent cognitive decline in aging individuals has already shown some success in preliminary population-based studies. This strategy is based on the fact that most persons at age 60 already show about 20 % less cerebral blood flow than at age 20, and adding an extra burden to this blood flow drop via conditions known to reduce brain perfusion will likely accelerate cognitive failure. However, by treating or managing vascular risk factors (Figs. 12.1, 12.2, and 12.3), considerable help in preventing the cerebral blood flow to drop to a critical level may be forestalled. But, what happens when treating or managing vascular risk factors does not work?

Low-level laser therapy, also known as photobiomodulation, is an emerging therapeutic approach in which cells or tissues are exposed to noninvasive, low-levels of **red-to-near-infrared light** (NIR). Quantum mechanical theory defines light energy as composed of photons whose energy depends only on the wavelength. Therefore, the energy of light depends mainly on the number of photons emitted and on their wavelength. Photons delivered to the brain tissue using NIR wavelengths are absorbed by the mitochondria in brain neurons and interact with cytochrome oxidase which is the primary photoacceptor for the NIR range in mammalian cells [1]. The energy generated by NIR is energy the neuron can use, especially if the cell fuel ATP synthesis is underproduced from, for example, damage to the mitochondrial DNA or poor blood flow to the brain, as it occurs in age-dependent cognitive decline. In principle, when underproduced ATP in hypoxic cells threatens neuronal survival, increasing ATP production in the mitochondria by stimulating cytochrome oxidase may provide an effective high-energy rescue of cellular respiration, oxygenation, and neuronal function. A secondary benefit of increasing mitochondrial cytochrome oxidase with NIR photons is that nitric oxide is released and diffused outside the neuronal cell wall, creating increased local blood flow via vasodilation. This extra-generated local

J.C. de la Torre, *Alzheimer's Turning Point*,
DOI 10.1007/978-3-319-34057-9_18

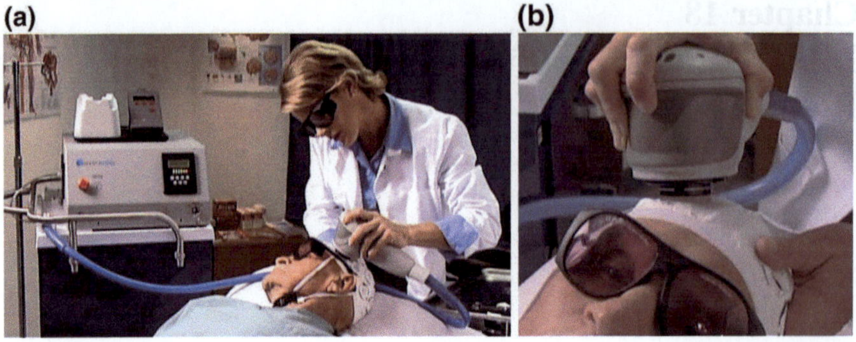

Fig. 18.1 Procedure for transcranial near-infrared laser therapy (NILT). Noninvasive application of NILT requires a wavelength and power density to be chosen dependent on the equipment being used, the duration of NILT application and anatomic location where near-infrared laser irradiation is used. NILT delivers laser photons that stimulate cytochrome oxidase in mitochondria within neurons, a process that drives the synthesis of the cell energy fuel ATP. The added ATP may counteract the neuronal energy crisis generated by chronic brain hypoperfusion that can trigger severe cognitive impairment and possibly prevent the clinical onset of Alzheimer's. Note that protective goggles are worn by patient and therapist. Courtesy of Photothera, Inc. Carlsbad, California (see text for details)

blood flow will enhance glucose and oxygen delivery to the compromised neurons creating more ATP.

NIR wavelengths (630–1000 nm) using low-energy lasers or light-emitting diode arrays have been reported to be able to manipulate a variety of biological processes in both cell cultured neurons [1] and animal models [2].

NIR light therapy wavelengths of 800–900 nm can penetrate the human scalp and skull for several centimeters to activate cytochrome c oxidase and mitochondrial synthesis of ATP [3] (Fig. 18.1). Experimental animal studies have reported that NIR can improve memory and reduce damage following traumatic brain injury in mice as well as enhance recovery from stroke in humans [4]. Other studies have shown that NIR can promote muscle repair, improve wound healing, and stimulate angiogenesis.

Normally, cell energy is extracted from foods we eat in a process where glucose is broken down into simpler molecules by the cell in a reaction called glycolysis. These simpler molecules are then fed to mitochondria within the cell and together with oxygen, and cytochrome oxidase is made which then produces ATP, the universal energy fuel for all cells.

Although photobiomodulation technology is still in its infancy of development, it has already been used in improving the recovery of ischemic injury from heart attacks, traumatic brain injuries, wounds, and in slowing down the degeneration of injured optic nerves [5] (Fig. 18.2).

A study led by **Dr. Francisco Gonzalez-Lima**, a professor of neuroscience at the University of Texas in Austin, provided the first study showing that transcranial laser photobiomodulation on the forehead of healthy human volunteers improved

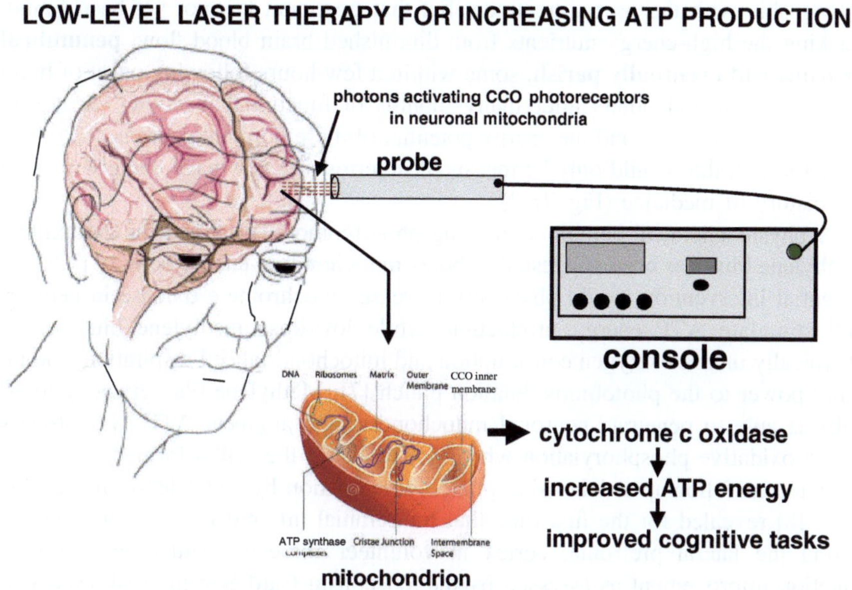

Fig. 18.2 Schematic diagram showing a transcranial noninvasive technique for increasing the energy fuel ATP in neurons of the prefrontal cortex using near-infrared laser (NIR) photons. A probe in contact with the forehead (and possibly anywhere on the bare scalp) dispenses low-level laser therapy (LLLT) dose housed in a portable console. Absorption of NIR photons can activate mitochondrial cytochrome c oxidase (CCO) in mitochondrial photoreceptors located in the inner membrane to produce an increase in cellular respiration and ATP energy. Pilot data in human volunteers indicate that executive function enhancement results from using this technique. It is theorized that increasing ATP synthesis by LLLT could be helpful in improving or preventing cognitive impairment, particularly in the elderly (see text for details)

memory tasks and measures of memory retrieval latency as compared to a placebo group receiving no photobiomodulation [6]. Laser exposure in the study by Gonzalez-Lima produces virtually no heat and no physical damage at the low power level used. Gonzalez-Lima also pointed out [6] the possible neuroprotective effect photobiomodulation might have in preventing a neuronal energy crisis such as the one that develops during the early stages of Alzheimer's. This would be tantamount to preventing Alzheimer's at a very early stage prior to the development of cognitive deterioration.

Photobiomodulation could also be considered as a neurorestorative treatment to reverse brain damaged neurons that survived the injury [1]. If successful, this could possibility revolutionize brain disorders, including dementia where penumbral neurons could be rescued from their physiological stupor and inactivity. Penumbral neurons are known to remain structurally undamaged following physical or physiological brain damage, but they do not function normally and appear more as "comatose" to any stimulating signal. Evidence is available that penumbral neurons are in a hypoperfused but not irreparable state. Such data suggest they are salvageable with the proper treatment but that treatment has not yet been found. Lacking the high-energy nutrients from diminished brain blood flow, **penumbral neurons will eventually perish**, some within a few hours following onset of brain hypoperfusion and others after many months of functional inactivity. Restoring cytochrome c oxidase and the energy potential of these dormant neurons could be a breakthrough that would only be measured in terms of a new page being written in the history of medicine (Fig. 18.2).

A physicochemical combination using photobiomodulation and the antioxidant methylene blue has been suggested to boost mitochondrial energy activity [7]. This potential intervention would physically increase cytochrome c oxidase in neurons and stimulate ATP energy production, while low-dose methylene blue would chemically increase oxygen consumption and mitochondrial cell respiration, adding more power to the photobiomodulation punch [7]. Methylene blue given systemically is able to penetrate neuronal mitochondria and augment ATP by a process called oxidative phosphorylation whereby energy for the cell is formed.

A recent followup article using photobiomodulation by Gonzalez-Lima and his team [8] revealed for the first time that transcranial infrared laser stimulation targeting the lateral prefrontal cortex in volunteer subjects resulted in executive function improvement as assessed by the Wisconsin Card Sorting Task (WCST). The WCST is a neuropsychological test that is considered the gold standard in evaluating cognitive functions such as working memory tasks and abstract reasoning ability when the prefrontal cortex is activated. The transcranial infrared laser stimulation was done in a single 8-min session. This intriguing study adds further evidence that photobiomodulation may become a useful tool one day involving conditions such as dementia by promoting mitochondrial energy in the form of ATP and protection from neurodegenerative disease.

Some experiments have reported that photobiomodulation can rescue neurons that have been functionally silenced with toxic chemicals by the ability of near-infrared light to raise the levels of the energy promoter cytochrome c oxidase inside these neurons [1]. If this and other studies are confirmed, photobiomodulation will be considered as a primary treatment for disorders that affect cellular metabolism and mitochondrial energy production in the brain.

As a novel technology still in its infancy, the questions raised by the use of transcranial infrared laser stimulation are many. For example, how long will mitochondrial production of ATP last after stimulation with transcranial infrared laser stimulation? What optimal wavelength and timing of the applied transcranial infrared laser stimulation will promote the best results to the cognitively impaired patient? How often will patients with cognitive issues need to be treated with transcranial infrared laser stimulation? What anatomic locations will transcranial infrared laser stimulation best enhance cognitive functions? How will cognitive improvement be measured in transcranial infrared laser stimulation-treated patients? Will photobiomodulation help people with mild cognitive impairment? The answer to these and other key questions will need to be addressed before general application of photobiomodulation is clinically applied to patients suffering or at risk of cognitive deterioration.

If one could roll the clock forward 50 years, it may not be difficult to imagine that photobiomodulation and the use of NIR or much more advanced technology based on low laser light therapy may be a standard tool in brain centers dedicated to increasing memory function and rehabilitating other cognitive defects brought on by aging, trauma, or disease. These laser devices might even be available for home use, given a set of instructions and safety features that would make it nearly impossible to misuse.

The journey for the medical application of photobiomodulation in brain repair, in restoration of lost neuronal function, and in improvement of cognitive impairment has begun. This journey is an exciting one because of the as yet unimaginable variety of uses photobiomodulation could have in people with brain damage, especially those who face the onslaught of dementia.

Potential Pharmacotherapy to Enhance Cerebral Blood Flow

A first principle in biology states that blood flow is the lifeline of every organism with a circulatory system—No blood flow, no life. Critical blood loss, heart attacks, and ischemic strokes can kill because blood flow to keep the brain alive is insufficient. But, what happens when cerebral blood flow is only moderately diminished as we age? This is a topic that has not been studied as much as it merits. The reason this topic is of critical importance is because substantial evidence now supports the precept that chronic brain hypoperfusion during advanced aging induces cognitive failure over time and eventual conversion to Alzheimer's pathology [9]. If chronic brain hypoperfusion can be controlled or reversed by interventions that directly elevate cerebral blood flow, it may be possible to prevent cognitive decline before major mental dysfunction is manifested.

There are no FDA-approved pharmacological agents in the USA that can increase blood flow to the brain in a sustained and sufficient manner or that can reverse brain ischemia.

Many pharmacological agents can raise cerebral blood flow, but these act mildly and for short durations. The key to successful treatment of reversing chronic brain hypoperfusion needs to address 3 major issues: (a) a chemical agent that is **safe**, (b) a chemical agent that can counterbalance **brain hypoperfusion**, and (c) a chemical agent that can **sustain** its action for many hours.

Increasing cerebral blood flow has always been a problem in the field of pharmacotherapy because the 3 major issues listed above could rarely be met by drug development. Consequently, a variety of drugs have been clinically tested in Alzheimer's patients with only mild and unsustainable effects including nonsteroidal anti-inflammatory drugs (NSAIDs), aspirin, vitamin E, physostigmine, cholinergics, estrogen, and vinpocetine. These medicines have a brief and unsustainable action on cerebral blood flow. Moreover, except for compounds that suppress the activity of carbonic anhydrase, the effect of these drugs on raising brain perfusion is trivial.

Acetazolamide, a well-known carbonic anhydrase inhibitor, has been used medically to increase blood flow in glaucoma and lower intraocular pressure. Acetazolamide has also been used orally to increase blood flow in mildly demented geriatric patients but with little success [10]. The effect of carbonic anhydrase inhibitors in increasing cerebral blood flow may be explained by their vasodilatory action and by decreasing brain pH.

As a strong vasodilator, acetazolamide has been used to evaluate cerebrovascular resistance. When acetazolamide is given intravenously, cerebral blood flow rises within 15 min 30–60 % above baseline in healthy subjects. Vascular resistance, which is a marker for reduced blood flow in the brain, can be calculated by subtracting baseline cerebral blood flow (taken before acetazolamide), from the post-acetazolamide cerebral blood flow. When acetazolamide does not increase cerebral blood flow above baseline flow, cerebrovascular resistance is compromised and brain ischemia is likely present from either emboli or hypoperfusion. The **acetazolamide challenge test** (as it is called) has also been used to detect hemodynamic abnormalities in patients with mild memory deficits [11], but strangely, it is not a standard tool for detecting cognitive impairment due to chronic brain hypoperfusion. The reason for this is unclear.

Several novel experimental drugs in the process of development have been reported that it could be useful to the problem of sustaining significant cerebral blood flow increase while providing safe limits for long-term use.

Nitric oxide is an important intercellular messenger in brain hemodynamics. Within the endothelial cells that line the inside wall of arterioles, nitric oxide is produced by endothelial nitric oxide synthase from its precursor, L-arginine. Nitric oxide has been shown to be an important regulator of local cerebral blood flow by its effect on the relaxation of a muscular layer of cells on the outside wall of arterioles (called vascular smooth muscle cells) which make vessels dilate and allow more blood to flow. Cerebral blood flow is impaired when nitric oxide production from endothelial cells becomes dysfunctional, for example, from distortions in brain small vessels and hemodynamic changes during advanced aging.

Compounds that activate endothelial nitric oxide synthase levels tend to increase cerebral blood flow, and this has been a therapeutic target in the attempt to elevate blood flow to the brain when needed. This is no easy task because endothelial nitric oxide synthase is involved in panoply of actions involving cardiovascular homeostasis, systemic blood pressure, the production of blood vessels (angiogenesis), and platelet clumping inside vessels. One agent that has been reported to increase endothelial nitric oxide synthase without harmful consequences is *Allium sativum*, otherwise known as common garlic [12]. This action may be due to garlic's ability to reduce blood pressure and prevent platelets from clumping within blood vessels where they can slow down flow. There is however, no evidence that garlic consumption is useful in elevating cerebral blood flow in humans.

Other ideas to increase endothelial nitric oxide synthase have been put forward such as using precursors of nitric oxide or compounds that may release nitric oxide into the circulation. Nitric oxide precursors can be found in many foods that make up the Mediterranean diet, fruits, nuts, grains, vegetables, fish, and olive oil in foods instead of salt. Skipping rope in the physically able is not only a good exercise in itself, but some theoretical assumptions suggest that it may activate endothelial nitric oxide synthase and nitric oxide production. At the present time, no drug has been found that can safely be used to manipulate nitric oxide production in the brain, but there is hope that research will discover a way to selectively do this in the near future.

Rho Kinase Inhibitors

Rho kinase (ROCK) is a serine–threonine enzyme that regulates, among other things, smooth muscle cell contraction, an action that will lower cerebral blood flow. Rock has been shown to play a role in cardiovascular and cerebrovascular blood vessel spasm, a condition that will severely limit local blood flow. ROCK is known to be activated by the presence of hypoxia and ischemia, two serious factors that will reduce cerebral blood flow by vasoconstriction. A large body of evidence has now been obtained regarding the important functions of ROCKs in vascular physiology, particularly in vascular smooth muscle cells that regulate cerebral blood flow. The major part of available data regarding ROCK-dependent functions in vascular smooth muscle cells has been obtained by the use of both in vitro and in vivo using the pharmacological ROCK inhibitors fasudil and hydroxyfasudil [13]. However, like all pharmacological agents, these inhibitors have only a relative specificity. Therefore, it is important to mention that the involvement of ROCKs in a particular function can be firmly established **only** when pharmacological data are supplemented by molecular analyses. ROCK is recognized as a major regulator of smooth muscle cell contraction but has also been demonstrated to be critical in controlling vascular smooth muscle cell migration, proliferation, survival, and programmed cell death, called *apoptosis*.

ROCK has also been implicated in lessening red blood cell deformability, a condition that can plug blood flow in cerebral capillaries where erythrocytes need to deform in order to pass through tight capillary lumens and bring oxygen to meet tissue needs. ROCK inhibitors given experimentally to animals appear to reverse the poor deformability of red blood cells and are being eyed as potential therapy for human conditions where hypoxia and ischemia are involved [14].

Reduced cerebral blood flow is associated with enhanced ROCK activity and with reduced endothelial nitric oxide synthase expression. When the selective ROCK inhibitors fasudil and hydroxyfasudil were used in a mice ischemic brain model, endothelial nitric oxide synthase expression and nitric oxide production became elevated, leading to an increase in cerebral blood flow in both hypoperfused and non-ischemic brain regions [15]. The increase in cerebral blood flow was shown to correlate well with the endothelial nitric oxide synthase increase [15]. The effect of fasudil in reversing ischemia has also been observed in the heart. Fasudil is used clinically as a therapeutic agent against cerebral vasospasm in Japan but is not approved in the USA.

The problem of creating new blood vessels in Alzheimer's brains that could transport nutrients to areas where capillaries are pathologically dysfunctional may require techniques that promote vasculogenesis. This process involves either embryonic new blood vessel formation, angiogenesis from preexisting damaged vasculature or neoangiogenesis, derived from adult new vessel formation.

Neoangiogenesis is an indirect approach to increase blood flow in the brain from the formation of new capillaries and blood vessels that were not present in the tissue before. It differs from angiogenesis where blood vessels are formed from a vascular system already present. Neoangiogenesis occurs mainly in embryonic development, but it is also present in wound sites during healing and is a feature of tumor growth.

Neoangiogenesis has been obtained experimentally with intravenous injections of vascular endothelial growth factor (VEGF) introduced into ischemic brain tissue [16]. VEGF is a powerful protein located in most cells that is able to stimulate angiogenesis and has been used experimentally for therapeutic effects in ischemia.

Since VEGF does not cross the blood–brain barrier easily, angiogenic gene therapy may offer the promise of delivering this protein directly into the brain. The technique would involve delivery of VEGF using plasmids, which are small, circular, and relatively inert DNA molecules. However, although plasmids could be injected directly into ischemic brain regions, thereby bypassing the blood–brain barrier, there is a possibility that they may be taken up only temporarily and in very few cells. Nevertheless, success in plasmid delivery of VEGF has been reported in animal models [17] and in patients with critical limb ischemia [18] and ischemic myocardium [19].

Despite vampire legends embracing sinister plots, blood transfusions from young people to people with Alzheimer's are being studied in earnest by a team at Stanford University. Heading the study is **Dr. Tony Wyss-Coray**, a professor of neurology at the medical school. The idea of rejuvenating old blood with younger one is not new having its origins from experiments done in the 1950s when old mice received young blood to repair bone cartilage. More recently, heterochronic parabiosis, as the young to old blood transfusion is called, has been used to reverse cardiac hypertrophy in old mice [20]. In that experiment, enlarged hearts from old mice exposed to young circulation for 4 weeks became markedly smaller than hearts in older mice not receiving this treatment.

ACE Inhibitors

Angiotensin-converting enzyme inhibitors (ACEI) are recognized as valuable therapeutic agents in controlling hypertension, coronary heart failure, and diabetic kidney disease. Angiotensin-converting enzyme (ACE) increases blood pressure by causing blood vessels to constrict and by degrading bradykinin, a powerful vasodilator in the renin–angiotensin system. The renin–angiotensin system is a blood pressure regulating hormone that can induce hypertension when it becomes abnormally active. Elevated ACE is found in many disorders that affect the liver, the lung, the kidney, the thyroid, the spleen, and the heart, all of which can directly or indirectly affect the brain.

Inhibition of ACE is a lifelong treatment for hypertension, and studies have found that incidental effects of ACEI seem to improve cognitive function and slow down cognitive decline in people with mild cognitive decline [21]. Some studies using ACEI in Alzheimer's patients have also reported a slower rate of cognitive decline. The reason for these improvements in cognition may be due to the ability of ACEI to produce vasodilation and improve blood flow to the brain in affected patients.

There is still much controversy surrounding the use of ACEI to improve cognitive function, but any agent that can increase blood flow to the brain in elderly individuals with hypertension or heart failure should be carefully evaluated in randomized controlled studies.

Sirtuins

Sirtuins (SIRT) are a family of proteins that regulate a wide variety of cellular processes, among them, mitochondrial energy metabolism and function. Understanding the physiological mechanisms that govern SIRT may provide a novel approach to control energy metabolic problems associated with aging. So far, 7 mammalian homologues of SIRT have been identified. Sirtuins 1-7 (SIRT1-7) belong to the class of enzymes which are dependent on NAD^+ for activity. NAD^+ (nicotinamide adenine dinucleotide) is a coenzyme found in all living cells which converts to NADH and together with molecular oxygen produces chemical energy in cells (see also Chap. 10, Glycolysis).

Research studies of SIRT support the notion that manipulating SIRT1 activity might result in protection from a host of metabolic derangements brought on by aging and hypoxia.

SIRT1 activation studies show that SIRT1 could be a novel therapeutic target in the management of metabolic energy disease and has prompted considerable research to identify other activators of SIRT1. For example, resveratrol, a substance found in red wine, has been shown to improve mitochondrial biogenesis and function by enhancing NAD^+ levels and promoting SIRT1 activation in mice. Based on this evidence, resveratrol could be useful in the treatment or prevention of age-related decline in cardiac function and neurometabolic damage, but more research is needed to verify this assumption.

Other SIRT activators such as the flavonoids fisetin and quercetin have been extracted from apples, strawberries, and grapes as well as red wine. The potential benefits of SIRT activators in aiding the aging process from diseases that will threaten cognitive function remain largely unexplored. However, the appeal of SIRT activators is that they can be found in healthy foods or red wine and consequently have a low risk of side effects for the elderly individual when properly given as supplements to their diet. Moreover, the door has recently opened to countless new synthetic SIRT activators that are considerably more potent than resveratrol. Two such synthetic SIRT1 activators SIRT-2104 and SIRT-1720 are presently undergoing clinical studies. These were designed to slow down the aging process and delay a number of age-related diseases in humans, such as inflammatory processes and cardiovascular disorders, two precursors of Alzheimer's dementia. It is conceivable that SIRT1 activators could constitute a novel class of drugs that could benefit the elderly and perhaps even middle-aged individuals, a concept that needs to be tested in randomized controlled human trials.

An intriguing study on the role of **fibrinogen** in Alzheimer's was reported by **Dr. Sydney Strickland** and his team from The Rockefeller University in New York city [22]. Fibrinogen is a protein in the blood plasma produced by the liver that has a key action in stopping blood loss from damaged tissue. Fibrinogen is converted to fibrin by the action of thrombin to induce coagulation of the blood during bleeding. Strickland and his team found that fibrinogen (or fibrin) deposits with Abeta in the brain tissue and blood vessels of Alzheimer's patients. These investigators have suggested that fibrinogen has a role in the pathophysiology of Alzheimer's dementia. The association between Abeta and fibrinogen led Strickland's team to propose that it could explain the cerebral hypoperfusion and subsequent neuronal pathology typically seen in Alzheimer's. It might also promote stroke which is often seen before and after Alzheimer's [22]. It is not clear why or how fibrinogen levels might increase in the plasma to cause deposition in the brain of Alzheimer's victims together with Abeta, but there could be a treatment that may prevent this potentially prothrombotic picture.

For example, factor Xa is an enzyme synthesized in the liver whose main function is blood clot formation in the coagulation process. Factor Xa direct inhibitors of the coagulation process including fibrinogen and thrombin are being used clinically in the USA to prevent strokes in the three million people with atrial fibrillation and those with a tendency to produce deep vein clots in their legs that can travel to the lung and cause respiratory failure. These **new oral anticoagulants (NOACS)** such as **apixaban** (Elequis) and **rivaroxaban** (Xarelto) have a low risk of hemorrhage and have become the preferred treatment for preventing thrombotic events.

Could NOACS be useful in preventing fibrinogen accumulation in the brain of healthy people who show a risk to Alzheimer's disease? This is an issue that has not been explored nor has there been any study so far on the use of NOACS in preventing Alzheimer's despite a large population of elderly people who are now taking these medications for a variety of disorders.

The rationale for using pharmacotherapy to increase cerebral blood flow to the brain is mounting due to clinical evidence that brain hypoperfusion is associated with mild cognitive impairment and is likely the cause of this condition. Cerebral hypoperfusion secondary to vascular risk factors during aging delivers suboptimal amounts of glucose and oxygen to hypometabolic brain cells.

Hypoxia from brain hypoperfusion additionally blocks mitochondrial oxidative phosphorylation, thereby inhibiting the synthesis of the energy cell fuel ATP and the reoxidation of NADH. This process leads to a marked decrease in ATP and an increase in the NADH/NAD ratio, a marker of mitochondrial dysfunction.

This pathological process may be corrected by increasing blood flow to the brain using new pharmacological compounds now in development. However, increased blood flow to the brain can conceivably be bypassed using photobiomodulation (as discussed above) to stimulate mitochondrial ATP production directly, thus providing hypoperfused brain cells with the energy boost needed to maintain optimal neurometabolism and survival.

If further studies are consistent with the concept of photobiomodulation stimulating mitochondrial oxidative phosphorylation and ATP production, it could introduce a simple and efficient method of preventing or treating the dark specter of cognitive meltdown and dementia.

References

1. Wong-Riley MT, Bax X, Buchmann E, Whelan HT. Light-emitting diode treatment reverses the effect of tetradotoxin on cytochrome oxidase neurons. NeuroReport. 2001;12:3033–7.
2. Karu T. Mitochondrial mechanisms of **photobiomodulation** in context of new data about multiple roles of ATP. Photomed Laser Surg. 2010;28(2):159–60.
3. Pitzschke A, Lovisa B, Seydoux O. Red and NIR light dosimetry in the human deep brain. Phys Med Biol. 2015;60:2921–37.
4. Naeser MA, Zafonte R, Krengel MH, Martin PI, Frazier J, Hamblin MR, Knight JA, Meehan WP 3rd, Baker EH. Significant improvements in cognitive performance post-transcranial, red/near-infrared light-emitting diode treatments in chronic, mild traumatic brain injury: open-protocol study. J Neurotrauma. 2014;31(11):1008–17.
5. Desmet KD, Paz DA, Corry JJ, Eells JT, Wong-Riley MT, Henry MM, Buchmann EV, Connelly MP, Dovi JV, Liang HL, Henshel DS, Yeager RL, Millsap DS, Lim J, Gould LJ, Das R, Jett M, Hodgson BD, Margolis D, Whelan HT. Clinical and experimental applications of NIR-LED photobiomodulation. Photomed Laser Surg. 2006;24(2):121–8.
6. Gonzalez-Lima F, Barrett DW. Augmentation of cognitive brain functions with transcranial lasers. Front Syst Neurosci. 2014;8:36–42.
7. Gonzalez-Lima F, Auchter A. Protection against neurodegeneration with low-dose methylene blue and near-infrared light. Front Cell Neurosci. 2015;9:179.
8. Blanco NJ, Maddox WT, Gonzalez-Lima F. Improving executive function using transcranial laser stimulation. J Neuropsychol May 28, 2015 (in press).
9. Di Marco LY, Venneri A, Farkas E, Evans PC, Marzo A, Frangi AF. Vascular dysfunction in the pathogenesis of Alzheimer's disease—a review of endothelial-mediated mechanisms and ensuing vicious circles. Neurobiol Dis 2015;82:592-606.
10. Wyper DJ, McAlpine CJ, Jawad K, Jennett B. Effects of carbonic anhydrase inhibitors on cerebral blood flow in geriatric patients. J Neurol Neurosurg Psychiatry. 1976;39:885–9.
11. Boles Ponto LL, Magnotta V, Moser DJ, Duff KM, Schultz KM. Global cerebral blood flow in relation to cognitive performance and reserve in subjects with mild memory deficits. Mol Imaging Biol 2006;8:253–72.
12. Das I, Khan NS, Sooranna S. Potent activation of nitric oxide synthase by garlic: a basis for its therapeutic activation. Curr Med Res Opin. 1995;13:257–63.

13. Shin HK, Salomone S, Potts EM, Lee SW, Millican E, Noma K, et al. Rho-kinase inhibition acutely augments blood flow in focal cerebral ischemia via endothelial mechanisms. J Cereb Blood Flow Metab. 2007;27:998–1009.
14. Thuet KM, Bowles EA, Ellsworth ML, Sprague RS, Stephenson AH. The Rho kinase inhibitor Y-27632 increases erythrocyte deformability and low oxygen tension-induced ATP release. Am J Physiol Heart Circ Physiol. 2011;301:H1891–6.
15. Rikitake Y, Kim HH, Huang Z, Seto M, Yano K, Asano T, Moskowitz MA, Liao JK. Inhibition of Rho kinase (ROCK) leads to increased cerebral blood flow and stroke protection. Stroke. 2005;36(10):2251–7.
16. Tsunumi Y, Takeshita S, Chen D, Keaney M, Rossow S. Direct intramuscular gene transfer of naked DNA encoding vascular endothelial growth factor augments collateral development and tissue perfusion. Circulation. 1996;94:3281–90.
17. Baumgartner I, Pieczek A, Manor O, Blair R, Kearney M, Walsh K, Isner JM. Constitutive expression of phVEGF165 after intramuscular gene transfer promotes collateral vessel development in patients with critical limb ischemia. Circulation. 1998;97(12):1114–23.
18. Takeshita S, Zheng LP, Brogi E, Kearney M, Pu LQ, Bunting S, Ferrara N, Symes JF, Isner JM. Therapeutic angiogenesis. A single intraarterial bolus of vascular endothelial growth factor augments revascularization in a rabbit ischemic hind limb model. J Clin Invest. 1994;93 (2):662–70.
19. Losordo DW, Vale PR, Symes JF, Dunnington CH, Esakof DD, Maysky M, Ashare AB, Lathi K, Isner JM. Gene therapy for myocardial angiogenesis: initial clinical results with direct myocardial injection of phVEGF165 as sole therapy for myocardial ischemia. Circulation. 1998;98(25):2800-4.
20. Loffredo FS, Steinhauser ML, Jay SM, et al. Growth differentiation factor 11 is a circulating factor that reverses age-related cardiac hypertrophy. Cell. 2013;153(4):828–39.
21. Solfrizzi V, Scafato E, Frisardi V, Seripa D, Logroscino G, Kehoe PG, Imbimbo BP, Baldereschi M, Crepaldi G, Di Carlo A, Galluzzo L, Gandin C, Inzitari D, Maggi S, Pilotto A, Panza F; Italian Longitudinal Study on Aging Working Group. Angiotensin-converting enzyme inhibitors and incidence of mild cognitive impairment. The Italian Longitudinal Study on Aging. Age (Dordr). 2013;35(2):441–53.
22. Cortes-Canteli M, Mattei L, Richards AT, Norris EH, Strickland S. Fibrin deposited in the Alzheimer's disease brain promotes neuronal degeneration. Neurobiol Aging. 2015;36 (2):608–17.

Chapter 19
Great Expectations

What to Expect from Alzheimer's Research

Like the powerful intense novel by Charles Dickens who describes with wit and compassion the long road to travel before good triumphs over evil, *Great Expectations* are two words that can serve to describe the long road to limiting Alzheimer's pathogenesis. Let us look at the problem in a nutshell.

The projected increase of 14 million people affected with Alzheimer's by 2050 in the USA is appalling. With about 5 million people now diagnosed with this dementia in the USA, it is anticipated that the number of new cases will consume more that 30 % of the Medicare and Medicaid budget at an annual cost of over 1 trillion dollars.

What is the present fate of a person diagnosed with mild memory loss who may be at risk of contracting this terrible dementia?

It is important for the reader to grasp how Alzheimer's dementia starts, why it continues relentlessly, why there has been no clinical progress in the last 100 years, how inadequate research funding impacts the disease, how greedy pharmaceuticals manipulate what research will be done, how false premises and money-making schemes delay research progress, what really causes Alzheimer's and how immediate steps in the development of **Heart-Brain Clinics** can markedly slow down or prevent the exploding number of new cases expected every year [1]. These questions are reviewed in this book and while no earth-shaking answers are immediately palpable, the problem is at least recognized, meticulously analyzed, and tentative strategies are offered to replace political complacency and clinical stagnation.

Now is the time to lift a decade's long comatose research out of its dark hole and replace it with new knowledge and new thinking that can jolt the field of Alzheimer's disease into a dramatic turning point from its present medical quagmire.

One important message that is carefully reviewed in this book is that cognitive or severe memory loss during advanced aging is not a normal or inevitable part of

J.C. de la Torre, *Alzheimer's Turning Point*,
DOI 10.1007/978-3-319-34057-9_19

growing old but merely reflects symptoms that can be averted by a technology that can be put in place in the next few years.

As will be seen, this clinical expectation has resulted from a substantial number of recent studies indicating that **identifying and detecting the major risk factors associated with Alzheimer's before its onset can provide early preclinical interventions to substantially stave off progressive cognitive failure**.

The long road to controlling Alzheimer's disease will not come with the discovery of a magic bullet that will cure the disease. **Anyone who tells you Alzheimer's can be cured is either a liar or an idiot. The liar will take your money and run, the idiot will babble nonsense without facing the facts**. The facts are, as reviewed in Chap. 13, (Bad aging) is that we cannot resuscitate, replace or otherwise modify the billions of dead neurons and their synapses that hold our memories and participate in our cognitive abilities.

To replace this population of dead neurons (even if it were possible), would be akin to building a new mind, actually a new being, using the old body as a shell. Let us be clear about nerve cell regeneration. Other brain neurons **not responsible for cognitive function** could someday be replaceable, and it is entirely possible that restoration of lost motor or sensory function can be achieved in the future. That is the good news for the victims of stroke, multiple sclerosis, spinal cord paralysis, viral encephalitis, amyotrophic lateral sclerosis, Parkinson's disease, and a host of other motor–sensory disorders. But not Alzheimer's or any other condition that wipes out cognitive function and makes us who we are.

How to Move Forward in Alzheimer's Research

The human brain is an exceptional organ. Billions of brain cells and trillions of synaptic connections allow us to speak, hear, think, remember, feel emotions, and experience consciousness. One neuron may communicate with thousands of other neurons by means of electrochemical messengers that whiz at flashing speeds from synapse to synapse to activate a mental, or movement or feeling response. Healthy communication among neurons is fundamental to being aware. Not being aware means not experiencing our environment, other people, or ourselves. This is what happens during Alzheimer's. When the neural connections in cognitive regions die or are broken, we become unaware of things even though we may still see, hear, move, and otherwise appear externally "normal." Neuronal death and their broken connections in dementia are irreversible and cannot be fixed.

We are left with only one option, to prevent brain cell death for as long as possible. I am convinced this will be practical to do one day. Its eventual realization in the USA will depend on a large extent on the political mood of the people in demanding adequate funding from Congress and employing a lobbying group to accelerate, implement, and realize a prevention strategy such as I have outlined in Chap. 16, under the section **Heart-Brain Clinics can lower the number of new Alzheimer's cases**.

A support for a newly organized Alzheimer's advocacy group that can jolt Congress into funding Heart-Brain Clinics and the staff required to make it work can learn from the AIDS/HIV advocacy groups that had great success in squeezing large sums of funding money from the National Institutes of Health (NIH) in the early 1980s. AIDS activists rewrote the book on how to persuade Congress and pressure President Reagan to increase research funding for AIDS patients. This, together with repeated public protests paid off, with AIDS now getting more money per patient than any other disease. Curiously, AIDS research gets **3 billion annually from NIH while Alzheimer's disease gets under** 600 million despite the fact that Alzheimer's affects five times more the number of people when compared to AIDS.

A report from **Dr. Rachel Best** [2], a sociologist from the University of Michigan, wrote in the American Sociological Review in 2012, that lobbying provides a way for taxpayers to influence how the NIH allocates its funding priorities and this tactic often works. These health-advocacy groups have one major purpose, to lobby Washington for more research funding for their favorite disease. The advocacy group has grown in size over the last two decades and now support over 50 diseases. Best also found that the more the lobbying, the more the research dollars could be obtained and **surprisingly, she noted that the diseases that affect the most people, such as Alzheimer's, do not often get the strongest advocacy support**. One problem that follows successful lobbying for a favorite disease is that NIH funding cannot be stretched beyond its fixed budget unless it is substantially increased by Congress. Consequently, getting NIH to increase its funding for one disease means that another disease will get less money than originally allocated. This results in what is known as the "Matthew Effect," a reverse-Robin Hood message taken from the Gospel of Matthew that states: **"For him that hath, shall be given in abundance, but from him that hath not, shall be taken away."**

There are a number of Alzheimer's support groups in most states that provide information and caregiver advice. The main one, which also provides limited funding for Alzheimer's research projects, is the Alzheimer's Association, located in Chicago. However, this nonprofit organization has no clear strategy on how to whip dementia and most of its time is spent in seeking public donations. Its clout in Congress is minimal as evidenced by the meager funding Alzheimer's research receives annually from NIH.

Although Alzheimer's research funding deserves to markedly increase its allocation of $585 million by NIH for fiscal year 2017, the establishment of Heart-Brain Clinics dedicated to the detection of Alzheimer's risk factors and their prevention or management should be the most compelling priority to defeat this mind-robbing disease.

Cancer Centers have proliferated in the USA in the last decades that have helped reduce the incidence of death from this second leading cause of mortality. For example, at many cancer centers, state-of-the-art diagnostic and genetic tools are used to detect cancerous cells before the formation of tumors. The results of the detection are then routinely examined by cancer specialists with the help of auxiliary personnel. These highly trained professionals can apply innovative treatments and delivery methods that often lead to life-saving measures. Moreover, treatment methods are generally "personalized" to tailor-fit the patient's cancer. Cancer specialists at Cancer Centers often work together and pool their resources in an attempt to bring the most appropriate and effective treatment to each patient. This protocol seems to work well and is one of the reasons why many cancers are now highly treatable or put into remission, for example, cancers that invade the lung, prostrate, breast, thyroid, skin, testicles, or blood.

These Cancer Centers did not exist many years ago, but they slowly flourished in demand for better prevention and treatments to put cancer into remission. Many of these Cancer Centers additionally provide research and teaching to their personnel to keep them updated for changes in their disciplines, a plan that easily extends to training other doctors and support staff. The most uplifting aspect of Cancer Centers is that their daily experience with many patients has led to rapid improvements and breakthroughs in their constant fight against cancer.

A model of Cancer Centers with some obvious modifications could be established in major cities where Heart-Brain Clinics could be set up with initial government help (see also Chap. 14). **Establishing Heart-Brain Clinics in major US cities should have a dramatic impact in lowering Alzheimer's prevalence because detection and prevention of cognitive decline is presently within our grasp**.

The present system of assessing, predicting, and preventing incipient Alzheimer's barely works. The system is based on a middle age or elderly patient's visit to a hospital, or a neurologist or a family practitioner complaining of memory problems. If the memory problems are not severe, the patient is often sent home with reassurance that Alzheimer's is not present. The problem with this approach is that once Alzheimer's disease is diagnosed, there is little that can be done medically to restore lost cognitive function or to prevent progression of the dementia.

This monotherapeutic approach is a lost opportunity to provide the patient with state-of-the-art diagnostics in the form of highly sensitive brain imaging to evaluate cerebral blood flow and metabolism, neurocognitive testing, and cardiovascular assessment, all of which pooled together can generally predict who is at risk of developing Alzheimer's. Once risk of cognitive dysfunction or mild cognitive impairment is detected, early interventions aimed at risk management and risk stratification can be quickly applied to the patient.

Many companies have marketed or are trying to market biomarkers for Alzheimer's disease. This is easy money to be made because most people with mild or moderate memory may want to know if dyscognitive symptoms will lead to dementia one day. Such biomarkers are not a good idea even if they are able to predict Alzheimer's unless a multidisciplinary clinical center, for example, a Heart-Brain Clinic, can determine risk stratification of the test result and offer a "personalized medical plan" to identify and treat the trigger of the memory difficulties.

At the present, there is no single test that can predict the development of Alzheimer's. Neuroimaging tests are constantly being perfected to evaluate healthy middle age or older patients who are at risk of developing Alzheimer's [3]. Within a short time, these neuroimaging tests may improve to the point where they will become the gold standard of Alzheimer's predictive diagnosis, much like a transthoracic echocardiogram can predict heart failure noninvasively.

I have great expectations that Heart-Brain Clinics or something comparable will be established when the socioeconomic and medical gravity of dementia and specifically Alzheimer's is recognized by politicians and private industry. I am not counting on drug makers because there is much less money to be made from prevention. Besides, pharmaceuticals in the USA want to create medicines that are consumed by people every day for whatever chronic ailment. It may be cynical to say this, but the more disease, the more money big pharma will make. Pharmaceuticals already control the market on drugs that can help **delay** Alzheimer's such as antihypertensives, anti-lipidemics, and anti-diabetes. If an inexpensive anti-aging pill is ever discovered, drug companies will clearly charge gouging prices to meet public demand.

Drug companies now have a profit margin of 20–30 % on their products, one of the biggest profit margins of any industry. Lisa Joldersma of the Pharmaceutical Research and Manufacturers of America (PhRMA), an advocacy group that represents drug companies in Washington DC, was recently quoted as saying "We are not going to have an Alzheimer's cure if we have arbitrary price controls." This incredibly naive comment violates two rules: first, it is highly unlikely that an Alzheimer's cure will ever be found despite the repetitive drumbeating by big pharma that a new breakthrough elixir is now being clinically tested. Second, drug discoveries by foreign pharmaceuticals do have price controls that do not inhibit their research and development of new drugs. In fact, many foreign pharmaceuticals charge a fraction of the price for American drugs sold abroad and are still able to make a handsome profit.

References

1. de la Torre JC. In-house heart-brain clinics to reduce Alzheimer's disease incidence. J Alzheimers Dis. 2014;42(Suppl 4):S431–42.
2. Best R. Disease politics and medical research funding. Am Sociol Rev. 2012;77:780–803.
3. Xekardaki A, Rodriguez C, Montandon ML, Toma S, Tombeur E, Herrmann FR, Zekry D, Lovblad KO, Barkhof B F, Giannakopoulos P, Haller S. Arterial spin labeling may contribute to the prediction of cognitive deterioration in healthy elderly individuals. Radiology 2014;274:490–9.

short time, these neuroimaging tests may improve to the point where they will become the "gold standard" of Alzheimer's predictive diagnosis, much like a transthoracic echocardiogram can predict heart failure noninvasively.

I have great expectations that Heart Brain Clinics or something comparable will be established when the real economic and medical gravity of dementia and specifically Alzheimer's is recognized by politicians and private industry. I am not [illegible]

[illegible]

[illegible]

Chapter 20
A Road to New Thinking

This book attempts to explore some of the key elements that transform a healthy and active mind into a world of darkness and lost awareness. This is the world of Alzheimer's disease. Since no cure is unlikely to be discovered in the future due to the destruction of the cognitive-related neural networks, Alzheimer's must be thwarted by targeting two fronts: first, by creating highly effective diagnostic centers (Heart-Brain Clinics, see Chap. 16) that can detect the earliest signs and risk factors prior to advanced cognitive impairment and potential transformation to Alzheimer onset and second, by applying current preventive interventions and developing new interventions based on the database collected from the clinical findings obtained in the diagnostic centers.

Consider a skyscraper that is on the verge of a horrific fire from a tiny lethal spark buried somewhere in the building that may go off at any time threatening the life of all its occupants. The current solution? Try and put out the fire once it has consumed most of the building and people in it. But, suppose for argument's sake, some intelligent force is created whose target is to find the spark before the building is set ablaze and roasts everyone in it?

In this analogy to the manner in which Alzheimer's disease is presently treated and managed, there is a total lack of vision on how to best contain the disease. Two critical issues need to be considered in the case of a make-believe fire that can start from a hidden spark. First, an initiative must be established to recognize the danger of dying from such a fire due to a hidden spark. This recognition promotes the development of clever devices that can detect the presence of such a spark and the knowledge of what the spark can lead to if it remains unacknowledged. Detecting the presence of the spark is only the first step in preventing the blazing fire. The next step is to construct a tool that can put out the spark as soon as it is discovered, thus preventing the start of the fire. If that tool for some reason does not shutdown the spark, it is prudent to have a backup plan using another device that targets small fires before they become a killing inferno.

This spark–fire analogy is essentially the plan I have proposed to contain Alzheimer's. This plan requires new thinking and a dedication by extracurricular forces, namely Congress, insurance companies, capital investments, and private

J.C. de la Torre, *Alzheimer's Turning Point*,
DOI 10.1007/978-3-319-34057-9_20

industry to do what is needed. We now know that the critical sparks for Alzheimer's development are the vascular risk factors to this dementia which are commonly encountered during advanced aging. There are detecting tools as reviewed in Chap. 15, which can readily identify these vascular risk factors. More importantly, detection and prediction of cognitive decline can provide an initial target for successfully treating these vascular risk factors that can lead to Alzheimer's or vascular dementia. This detection and prevention stage is best realized with centers, such as **Heart-Brain Clinics** (discussed in Chap. 16), specifically dedicated to fighting Alzheimer's at its earliest preclinical stage. Specialists in the fields of heart and brain pathophysiology are trained to recognize the vascular risk factors associated with cognitive decline would be at the forefront of this prevention program. When the cognitive dysfunction spark cannot be snuffed out, a backup system to manage the early stages of Alzheimer's should be devised that can ease the suffering and anguish of these patients and their close relatives.

The fundamental issue to recognize about Alzheimer's is that by all counts, the disease has become the most challenging medical problem of the 21st century. If allowed to continue, Alzheimer's will become unmanageable as it spreads from a present 5.3 million people affected in the USA to 14 million by 2050. The healthcare costs will skyrocket out of control from a present $155 billion to about $1 trillion in 2050, making this the most costly malady of all time even when compared to cancer and heart disease combined. After age 85, an increasing number of Americans are afflicted with this dementia. It is the 6th leading cause of death in the USA, a figure which will rise dramatically in the coming years from the present 700,000 dementia-related deaths. These statistics are not confined to the USA since other countries also predict similar societal costs to their healthcare systems and rising prevalence of dementia in the next few decades.

These sobering statistics should be a wake-up call to take action, not tomorrow or the next day but now. Surely, if terrorists managed to kill in one day the same number of people who die of Alzheimer's every day, Congress would hysterically jump into immediate action by whatever means necessary to stop the "massacre." Why not do it for Alzheimer's when it is within our grasp?

The establishment of Heart-Brain Clinics in major US cities as I have outlined in Chap. 16 could substantially reduce the number of new dementia cases. These clinics would operate in a similar fashion to Cancer Centers dedicated to detect, diagnose, and initiate a personalized therapeutic plan to control the cancer. Heart-Brain Clinics would be staffed with a multidisciplinary group of specialists thoroughly familiar with the three main systems affected by Alzheimer's disease: the psyche, the heart, and the brain. Heart-Brain Clinics would have the state-of-the-art diagnostic tools and the staff to identify individuals at high risk of developing cognitive decline, ideally before Alzheimer symptoms have appeared. A personalized plan of action based on the detection–diagnostic findings would then be applied to prevent further escalation of the pathology detected. Follow-up visits to the clinics would be critical to evaluate the personalized plan of prevention and apply any changes deemed necessary. There is now a consensus among a growing number of Alzheimer researchers that vascular risk factors to Alzheimer's should be the main target of prevention. These ubiquitous vascular risk factors are likely the most important therapeutic targets in the overall plan to prevent Alzheimer's, as reviewed in Chap. 12.

Heart-Brain Clinics staffed by neurologists, cardiologists, neuroradiologists, psychologists, and a support staff would replace the haphazard monotherapeutic approach currently used to diagnose Alzheimer's where the patient receives a cursory examination, a treatment of dubious value, and is sent home to cope with the dementia, assuming the diagnosis made is correct. This approach is fraught with missed diagnoses, medical errors, the absence of follow-up visits, and inability to personalize treatments or management of active or nascent cognitive deficits. **There must be a better way, and there is**. Funding for the establishment of Heart-Brain Clinics may appear as a luxury society cannot afford, but it is a necessity we cannot afford to ignore if we are to rescue a fast-growing elderly population that is presently doomed to die in one of Nature's most cruel fates. Many Level I trauma hospitals in the USA already have the equipment and testing tools necessary to detect preclinical signs of incipient Alzheimer's, but the staff at these hospitals lack the training and expertise to personalize a treatment or management plan and to promote follow-up visits by patients.

I have been very critical in this book of drugmakers and of investigators who knowingly or stupidly have led us down the garden path for decades and forced many others to join them or be left behind in their careers. There is no need here to point a finger or name names, most of us in the field know who these drugmakers and their paid consultants or favorite sons are, but there is a need to confront the failed concepts and false hope created by greed and indifference to the victims of dementia. Many drugmaker executives and distinguished researchers have relatives or friends who have been diagnosed with Alzheimer's or present a high risk to dementia. It would be an extraordinary insight to know what is in their minds when they doggedly pursue failed concepts and nostrums to reverse a brain already devastated by irreversible memory loss.

Pharmaceuticals begin to seduce future physicians at the start of medical school not from generosity or altruism, but out of a crass strategic plan to better sell their products in the future. This plan by big pharma continues after graduation as doctors are lavished with meals, speaking fees, paid conference attendance, and many other perks. Such friendly persuasion by offering legal bribes to doctors who prescribe medicines is generally not in the best interests of the patient [1].

The same applies when unfounded rhetoric replaces the scientific method, and truth is bent like a pretzel. The result is scientific stagnation and knee-jerk dogma that no longer seeks to search for truth. Something like this has happened to Alzheimer's clinical research, and the stranglehold created by an enduring but false paradigm has all but stopped all research initiatives and progress. After 100,000 clinical and scientific publications in the field of Alzheimer's in the last three decades, there is currently no hope, no effective treatment, and no incentive to know what causes this dementia. The truth hangs like an ornament on a Christmas tree to be admired but not really questioned.

This intellectual quagmire and the continuing coercive support by some subscribers to a failed Alzheimer hypothesis require a mini-scientific revolution to topple the trackless clinical outlook that has stuck like a bubo on our conscience and thinking.

We should recall Thomas Kuhn's aphorism concerning scientific revolutions:

> Because scientists are reasonable people, one or another argument will ultimately persuade many of them. But there is no single argument that can or should persuade them all. Rather than a single group conversion, what occurs is an increasing shift in the distribution of professional allegiances.

This book is not about forming allegiances to one theory over another. Like Kuhn, I believe the undecided scientists will find a way to get on the right research track, whatever that may be, not because it offers lucrative rewards but because it will lead to intellectual satisfaction. In my judgment, this is the main reason why some people become scientists. In rare cases, financial and intellectual satisfaction can be compatible and complimentary as seen by the development of a polio vaccine or the mapping of the human genome. More commonly, financial reward and intellectual integrity concerning a scientific endeavor are generally a choice to make.

This book briefly explores how a memory is created or destroyed. I am fascinated by the formation of memories and how they mold our personality and senses and more so, when they lead us to false impressions and the dark world of confusion and intellectual isolation. We know so little about this that it is almost tempting to ignore the layers of complexity involved in the memory domain and get on with something more substantive. However, my temptation to ignore the mind and its secrets is insubstantial because I am fully confident that we will know a great deal about it in times to come. For example, much of what I know and have read tells me that Cartesian dualism will never explain the chemical molecules that interact in a jelly-like medium called the brain and which provide us with thinking, consciousness, and awareness that can be easily erased by an insidious physical or physiological process. I do not pretend to know what I do not know but I feel intuitively that these chemical molecules darting about neural circuits will one day be dissected and analyzed in test tubes or mass spectrometers much like quarks, gluons, and other subatomic particles are now inspected. Until then, we need to toil with the tools at hand to try and make some sense of Nature's intentions.

We are on the cusp of helping millions of people around the planet to have a good chance of avoiding or delaying Alzheimer's disease. Considering my involvement with Alzheimer research for the last 25 years, I am convinced that this dementia can be significantly contained using the strategies described in this book. Other constructive ideas and initiatives will surely evolve as Heart-Brain Clinics create a database to assist in the process of healthy brain aging and push-back senescent neurodegeneration.

If history is a wise teacher of new ideas and forward thinking, I am confident that good medical research and practice will turn back the specter of dementia, not just for a lucky few who manage to escape the gauntlet of age-related maladies but for most people reaching the enviable elder stage of life. To do otherwise, is to ignore the panoply of human survival.

Reference

1. Brody H. The company we keep: why physicians should refuse to see pharmaceutical representatives. Ann Fam Med. 2005;3(1):82–5.

Index

J.C. de la Torre, *Alzheimer's Turning Point*,
DOI 10.1007/978-3-319-34057-9

MIX
Papier aus verantwortungsvollen Quellen
Paper from responsible sources
FSC® C105338

If you have any concerns about our products,
you can contact us on
ProductSafety@springernature.com

In case Publisher is established outside the EU,
the EU authorized representative is:
Springer Nature Customer Service Center GmbH
Europaplatz 3, 69115 Heidelberg, Germany

Printed by Libri Plureos GmbH
in Hamburg, Germany